FREE Test Taking Tips DVD Offer

To help us better serve you, we have developed a Test Taking Tips DVD that we would like to give you for FREE. **This DVD covers world-class test taking tips that you can use to be even more successful when you are taking your test.**

All that we ask is that you email us your feedback about your study guide. Please let us know what you thought about it – whether that is good, bad or indifferent.

To get your **FREE Test Taking Tips DVD**, email freedvd@studyguideteam.com with "FREE DVD" in the subject line and the following information in the body of the email:

    a. The title of your study guide.

    b. Your product rating on a scale of 1-5, with 5 being the highest rating.

    c. Your feedback about the study guide. What did you think of it?

    d. Your full name and shipping address to send your free DVD.

If you have any questions or concerns, please don't hesitate to contact us at freedvd@studyguideteam.com.

Thanks again!

ASVAB Study Guide 2020-2021

ASVAB Study Guide 2020 & 2021 and Practice Test Questions Book for the Armed Services Vocational Aptitude Battery Exam [Includes Detailed Answer Explanations]

Test Prep Books

Table of Contents

Quick Overview

As you draw closer to taking your exam, effective preparation becomes more and more important. Thankfully, you have this study guide to help you get ready. Use this guide to help keep your studying on track and refer to it often.

This study guide contains several key sections that will help you be successful on your exam. The guide contains tips for what you should do the night before and the day of the test. Also included are test-taking tips. Knowing the right information is not always enough. Many well-prepared test takers struggle with exams. These tips will help equip you to accurately read, assess, and answer test questions.

A large part of the guide is devoted to showing you what content to expect on the exam and to helping you better understand that content. In this guide are practice test questions so that you can see how well you have grasped the content. Then, answer explanations are provided so that you can understand why you missed certain questions.

Don't try to cram the night before you take your exam. This is not a wise strategy for a few reasons. First, your retention of the information will be low. Your time would be better used by reviewing information you already know rather than trying to learn a lot of new information. Second, you will likely become stressed as you try to gain a large amount of knowledge in a short amount of time. Third, you will be depriving yourself of sleep. So be sure to go to bed at a reasonable time the night before. Being well-rested helps you focus and remain calm.

Be sure to eat a substantial breakfast the morning of the exam. If you are taking the exam in the afternoon, be sure to have a good lunch as well. Being hungry is distracting and can make it difficult to focus. You have hopefully spent lots of time preparing for the exam. Don't let an empty stomach get in the way of success!

When travelling to the testing center, leave earlier than needed. That way, you have a buffer in case you experience any delays. This will help you remain calm and will keep you from missing your appointment time at the testing center.

Be sure to pace yourself during the exam. Don't try to rush through the exam. There is no need to risk performing poorly on the exam just so you can leave the testing center early. Allow yourself to use all of the allotted time if needed.

Remain positive while taking the exam even if you feel like you are performing poorly. Thinking about the content you should have mastered will not help you perform better on the exam.

Once the exam is complete, take some time to relax. Even if you feel that you need to take the exam again, you will be well served by some down time before you begin studying again. It's often easier to convince yourself to study if you know that it will come with a reward!

Test-Taking Strategies

1. Predicting the Answer

When you feel confident in your preparation for a multiple-choice test, try predicting the answer before reading the answer choices. This is especially useful on questions that test objective factual knowledge. By predicting the answer before reading the available choices, you eliminate the possibility that you will be distracted or led astray by an incorrect answer choice. You will feel more confident in your selection if you read the question, predict the answer, and then find your prediction among the answer choices. After using this strategy, be sure to still read all of the answer choices carefully and completely. If you feel unprepared, you should not attempt to predict the answers. This would be a waste of time and an opportunity for your mind to wander in the wrong direction.

2. Reading the Whole Question

Too often, test takers scan a multiple-choice question, recognize a few familiar words, and immediately jump to the answer choices. Test authors are aware of this common impatience, and they will sometimes prey upon it. For instance, a test author might subtly turn the question into a negative, or he or she might redirect the focus of the question right at the end. The only way to avoid falling into these traps is to read the entirety of the question carefully before reading the answer choices.

3. Looking for Wrong Answers

Long and complicated multiple-choice questions can be intimidating. One way to simplify a difficult multiple-choice question is to eliminate all of the answer choices that are clearly wrong. In most sets of answers, there will be at least one selection that can be dismissed right away. If the test is administered on paper, the test taker could draw a line through it to indicate that it may be ignored; otherwise, the test taker will have to perform this operation mentally or on scratch paper. In either case, once the obviously incorrect answers have been eliminated, the remaining choices may be considered. Sometimes identifying the clearly wrong answers will give the test taker some information about the correct answer. For instance, if one of the remaining answer choices is a direct opposite of one of the eliminated answer choices, it may well be the correct answer. The opposite of obviously wrong is obviously right! Of course, this is not always the case. Some answers are obviously incorrect simply because they are irrelevant to the question being asked. Still, identifying and eliminating some incorrect answer choices is a good way to simplify a multiple-choice question.

4. Don't Overanalyze

Anxious test takers often overanalyze questions. When you are nervous, your brain will often run wild, causing you to make associations and discover clues that don't actually exist. If you feel that this may be a problem for you, do whatever you can to slow down during the test. Try taking a deep breath or counting to ten. As you read and consider the question, restrict yourself to the particular words used by the author. Avoid thought tangents about what the author *really* meant, or what he or she was *trying* to say. The only things that matter on a multiple-choice test are the words that are actually in the question. You must avoid reading too much into a multiple-choice question, or supposing that the writer meant something other than what he or she wrote.

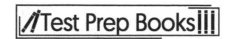

5. No Need for Panic

It is wise to learn as many strategies as possible before taking a multiple-choice test, but it is likely that you will come across a few questions for which you simply don't know the answer. In this situation, avoid panicking. Because most multiple-choice tests include dozens of questions, the relative value of a single wrong answer is small. As much as possible, you should compartmentalize each question on a multiple-choice test. In other words, you should not allow your feelings about one question to affect your success on the others. When you find a question that you either don't understand or don't know how to answer, just take a deep breath and do your best. Read the entire question slowly and carefully. Try rephrasing the question a couple of different ways. Then, read all of the answer choices carefully. After eliminating obviously wrong answers, make a selection and move on to the next question.

6. Confusing Answer Choices

When working on a difficult multiple-choice question, there may be a tendency to focus on the answer choices that are the easiest to understand. Many people, whether consciously or not, gravitate to the answer choices that require the least concentration, knowledge, and memory. This is a mistake. When you come across an answer choice that is confusing, you should give it extra attention. A question might be confusing because you do not know the subject matter to which it refers. If this is the case, don't eliminate the answer before you have affirmatively settled on another. When you come across an answer choice of this type, set it aside as you look at the remaining choices. If you can confidently assert that one of the other choices is correct, you can leave the confusing answer aside. Otherwise, you will need to take a moment to try to better understand the confusing answer choice. Rephrasing is one way to tease out the sense of a confusing answer choice.

7. Your First Instinct

Many people struggle with multiple-choice tests because they overthink the questions. If you have studied sufficiently for the test, you should be prepared to trust your first instinct once you have carefully and completely read the question and all of the answer choices. There is a great deal of research suggesting that the mind can come to the correct conclusion very quickly once it has obtained all of the relevant information. At times, it may seem to you as if your intuition is working faster even than your reasoning mind. This may in fact be true. The knowledge you obtain while studying may be retrieved from your subconscious before you have a chance to work out the associations that support it. Verify your instinct by working out the reasons that it should be trusted.

8. Key Words

Many test takers struggle with multiple-choice questions because they have poor reading comprehension skills. Quickly reading and understanding a multiple-choice question requires a mixture of skill and experience. To help with this, try jotting down a few key words and phrases on a piece of scrap paper. Doing this concentrates the process of reading and forces the mind to weigh the relative importance of the question's parts. In selecting words and phrases to write down, the test taker thinks about the question more deeply and carefully. This is especially true for multiple-choice questions that are preceded by a long prompt.

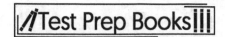

9. Subtle Negatives

One of the oldest tricks in the multiple-choice test writer's book is to subtly reverse the meaning of a question with a word like *not* or *except*. If you are not paying attention to each word in the question, you can easily be led astray by this trick. For instance, a common question format is, "Which of the following is…?" Obviously, if the question instead is, "Which of the following is not…?," then the answer will be quite different. Even worse, the test makers are aware of the potential for this mistake and will include one answer choice that would be correct if the question were not negated or reversed. A test taker who misses the reversal will find what he or she believes to be a correct answer and will be so confident that he or she will fail to reread the question and discover the original error. The only way to avoid this is to practice a wide variety of multiple-choice questions and to pay close attention to each and every word.

10. Reading Every Answer Choice

It may seem obvious, but you should always read every one of the answer choices! Too many test takers fall into the habit of scanning the question and assuming that they understand the question because they recognize a few key words. From there, they pick the first answer choice that answers the question they believe they have read. Test takers who read all of the answer choices might discover that one of the latter answer choices is actually *more* correct. Moreover, reading all of the answer choices can remind you of facts related to the question that can help you arrive at the correct answer. Sometimes, a misstatement or incorrect detail in one of the latter answer choices will trigger your memory of the subject and will enable you to find the right answer. Failing to read all of the answer choices is like not reading all of the items on a restaurant menu: you might miss out on the perfect choice.

11. Spot the Hedges

One of the keys to success on multiple-choice tests is paying close attention to every word. This is never truer than with words like almost, most, some, and sometimes. These words are called "hedges" because they indicate that a statement is not totally true or not true in every place and time. An absolute statement will contain no hedges, but in many subjects, the answers are not always straightforward or absolute. There are always exceptions to the rules in these subjects. For this reason, you should favor those multiple-choice questions that contain hedging language. The presence of qualifying words indicates that the author is taking special care with his or her words, which is certainly important when composing the right answer. After all, there are many ways to be wrong, but there is only one way to be right! For this reason, it is wise to avoid answers that are absolute when taking a multiple-choice test. An absolute answer is one that says things are either all one way or all another. They often include words like *every*, *always*, *best*, and *never*. If you are taking a multiple-choice test in a subject that doesn't lend itself to absolute answers, be on your guard if you see any of these words.

12. Long Answers

In many subject areas, the answers are not simple. As already mentioned, the right answer often requires hedges. Another common feature of the answers to a complex or subjective question are qualifying clauses, which are groups of words that subtly modify the meaning of the sentence. If the question or answer choice describes a rule to which there are exceptions or the subject matter is complicated, ambiguous, or confusing, the correct answer will require many words in order to be expressed clearly and accurately. In essence, you should not be deterred by answer choices that seem excessively long. Oftentimes, the author of the text will not be able to write the correct answer without

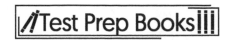

offering some qualifications and modifications. Your job is to read the answer choices thoroughly and completely and to select the one that most accurately and precisely answers the question.

13. Restating to Understand

Sometimes, a question on a multiple-choice test is difficult not because of what it asks but because of how it is written. If this is the case, restate the question or answer choice in different words. This process serves a couple of important purposes. First, it forces you to concentrate on the core of the question. In order to rephrase the question accurately, you have to understand it well. Rephrasing the question will concentrate your mind on the key words and ideas. Second, it will present the information to your mind in a fresh way. This process may trigger your memory and render some useful scrap of information picked up while studying.

14. True Statements

Sometimes an answer choice will be true in itself, but it does not answer the question. This is one of the main reasons why it is essential to read the question carefully and completely before proceeding to the answer choices. Too often, test takers skip ahead to the answer choices and look for true statements. Having found one of these, they are content to select it without reference to the question above. Obviously, this provides an easy way for test makers to play tricks. The savvy test taker will always read the entire question before turning to the answer choices. Then, having settled on a correct answer choice, he or she will refer to the original question and ensure that the selected answer is relevant. The mistake of choosing a correct-but-irrelevant answer choice is especially common on questions related to specific pieces of objective knowledge. A prepared test taker will have a wealth of factual knowledge at his or her disposal, and should not be careless in its application.

15. No Patterns

One of the more dangerous ideas that circulates about multiple-choice tests is that the correct answers tend to fall into patterns. These erroneous ideas range from a belief that B and C are the most common right answers, to the idea that an unprepared test-taker should answer "A-B-A-C-A-D-A-B-A." It cannot be emphasized enough that pattern-seeking of this type is exactly the WRONG way to approach a multiple-choice test. To begin with, it is highly unlikely that the test maker will plot the correct answers according to some predetermined pattern. The questions are scrambled and delivered in a random order. Furthermore, even if the test maker was following a pattern in the assignation of correct answers, there is no reason why the test taker would know which pattern he or she was using. Any attempt to discern a pattern in the answer choices is a waste of time and a distraction from the real work of taking the test. A test taker would be much better served by extra preparation before the test than by reliance on a pattern in the answers.

FREE DVD OFFER

Don't forget that doing well on your exam includes both understanding the test content and understanding how to use what you know to do well on the test. We offer a completely FREE Test Taking Tips DVD that covers world class test taking tips that you can use to be even more successful when you are taking your test.

All that we ask is that you email us your feedback about your study guide. To get your **FREE Test Taking Tips DVD**, email freedvd@studyguideteam.com with "FREE DVD" in the subject line and the following information in the body of the email:

- The title of your study guide.
- Your product rating on a scale of 1-5, with 5 being the highest rating.
- Your feedback about the study guide. What did you think of it?
- Your full name and shipping address to send your free DVD.

Introduction to the ASVAB Test

Function of the Armed Services Vocational Aptitude Battery (ASVAB) Test

The **Armed Services Vocational Aptitude Battery** (ASVAB) measures developed abilities and helps to predict academic and occupational success in the military. The ASVAB was first introduced in 1968, and over 40 million individuals have taken the exam since its introduction. Once a military applicant takes the ASVAB, military personnel use the score to assign an appropriate job in the military. The test also assesses whether an applicant is qualified to enlist in the military.

The test can be administered via paper and pencil (P&P-ASVAB) or via computer. The computer version – known as the CAT-ASVAB – was implemented after twenty years of evaluation and research. This was the first adaptive test battery to be administered as a high-stakes test. An adaptive test means the test adapts to an individual's ability level based on the answer to the previous question. A harder question will be administered if the previous question was answered correctly, and an easier question will be administered if the previous question was answered incorrectly. There is no entrance exam for the ASVAB.

Test Administration

The ASVAB is administered annually and is taken by over one million military applicants. Most applicants are high school or post-secondary students. Testing is conducted at a Military Entrance Processing Station (MEPS), or if this site is not available, the ASVAB can be taken at a satellite location known as a Military Entrance Test (MET) site. The ASVAB is computer-based at the MEPS but paper-based at most MET sites. The computer-based test takes about one-and-a-half hours for an average person to complete, while the paper and pencil version takes about three hours to complete. If a test taker needs to retake the test, he or she must wait at least a month before retaking. After the second retest, a test taker must wait six months to retest again. Scores may be used for enlistment purposes for up to two years after the date of testing.

Test Format

The paper and pencil test has 225 questions that are broken up into sections with time limits equaling a total of 149 minutes. There is no penalty for guessing on the P&P- ASVAB, and in fact, unanswered questions are marked as incorrect. Because the CAT-ASVAB is an adaptive test, the time limit may vary, but the computer-based version is usually shorter than the paper and pencil administration. There is a penalty for incorrect answers on the CAT-ASVAB, so it is ideal for a tester to answer as much as they can without randomly guessing. Unanswered questions also receive a penalty, which is found by calculating the score that would have been achieved had the test taker randomly guessed on unanswered questions. Both versions of the exam contain four domains: Verbal, Math, Science and Technical, and Spatial.

Below is a chart listing the content tests on the ASVAB and the domain they fall under. The chart lists the tests in the order in which they appear on the exam.

Test	Domain
General Science (GS)	Science/Technical
Arithmetic Reasoning (AR)	Math
Word Knowledge (WK)	Verbal
Paragraph Comprehension (PC)	Verbal
Mathematics Knowledge (MK)	Math
Electronics Information (EI)	Science/Technical
Auto Information (AI)	Science/Technical
Shop Information (SI)	Science/Technical
Mechanical Comprehension (MC)	Science/Technical
Assembling Objects (AO)	Spatial

Scoring

ASVAB scores are determined using an **Item Response Theory** (IRT) model, which tailors the test questions to the ability level of the test taker and enables scores to be presented on the same scale, even though there is variability in the questions attempted by each test taker. The model behind IRT is called 3PL, which represents difficulty, discrimination, and guessing, and is based on the likelihood that a test taker will provide a correct response for each specific question, given his or her performance on other questions. After the final ability estimate is measured, a standardized score is reached. Test takers receive standardized subtest scores and a composite score for the entire exam. Through a process called equating, scores on both versions of the ASVAB can be statistically and fairly compared.

Recent/Future Developments

There are future plans to continue to administer more of the CAT-ASVAB. Sites that currently have the P&P-ASVAB will eventually be only administering the CAT-ASVAB.

General Science

Geology

Geology is the study of the nature and composition of the rocks and materials that make up the Earth, how they were formed, and the physical and chemical processes that have changed Earth over time.

Earth can be imagined as a giant construction of billions of Lego blocks, and that these blocks represent different minerals. A **mineral** is a naturally occurring inorganic solid composed of certain chemical elements (or atoms) in a defined crystalline structure. When minerals are aggregated together with other minerals, **organic compounds** (carbon-containing remains of decomposed plant or animal matter), and/or **mineraloids** (minerals that lack a defined crystalline structure), rocks are formed. Rock types are classified based on their mechanism of formation and the materials of their compositions. The three fundamental classifications of rocks include:

- Sedimentary
- Igneous
- Metamorphic

Sedimentary rocks form at the Earth's surface (on land and in bodies of water) through deposition and cementation of fragments of other rocks, organic matter, and minerals. These materials, called sediment, are deposited and accumulate in layers called strata, which get pressed into a solid over time when more sediment settles on top. Sedimentary rocks are further classified as either clastic/detrital, biochemical, chemical, or other. **Clastic** or **detrital** rocks are composed of other inorganic rocks or organic particles, respectively. **Biochemical** rocks have an organic component, like coal, which is composed of decayed plant matter. **Chemical** rocks form from a solution containing dissolved materials that became supersaturated and minerals precipitate out of solution. Halite, or rock salt, is an example of a chemical sedimentary rock. Sedimentary rocks that do not fit into these types are categorized as "other." These rocks are from fragments formed by asteroid or comet impacts or from fragments of volcanic lava.

Igneous rocks are composed of molten material beneath the Earth's surface called **magma** and are classified based on where the magma cooled and solidified. They can be intrusive/plutonic, extrusive/volcanic, or hypabyssal. **Intrusive** or **plutonic** rocks, such as granite, form when magma cools slowly within or beneath the Earth's surface. Because they solidify slowly, these rocks tend to have a coarse grain, larger crystalline structure of their mineral constituents, and rough appearance. By contrast, extrusive/volcanic form from rapid cooling as magma escapes the Earth's surface as lava and have a smooth or fine-grained appearance, with tiny crystals or ones that are too small to see. A common example of an extrusive igneous rock is glassy obsidian. **Hypabyssal** rocks are formed at levels between intrusive and extrusive (just below the surface); they aren't nearly as common.

Metamorphic rocks form from the transformation of other rocks via a process called metamorphism. This transformation happens when existing rocks—sedimentary, igneous, or other metamorphic rocks— are subjected to significant heat and pressure, which causes physical and/or chemical changes. Based on their appearance, metamorphic rocks are classified as either foliated or non-foliated. **Foliated** rocks are layered or folded, which means they form from compression in one direction and result in visible layers or banding within the rock. Examples include gneiss or slate. **Non-foliated** rocks, such as marble, receive equal pressure from all directions and thus have a homogenous appearance.

It should be noted that classification is not always completely clear, and some rocks don't quite fit the criteria for one of these three categories, so they are sometimes lumped together in a category called "other rocks." A classic example is a fossil. A **fossil** is a rock formed from the remains or impression of dead plants or animals, but it doesn't fit into any biochemical class of rocks because fossils themselves are wholly composed of organic material and only formed under strict conditions, although they are commonly found within sedimentary rock.

Plate Tectonics

The theory of plate tectonics states that the Earth's superficial layer (the crust and upper mantle, together called the **lithosphere**) is a collection of variably-sized plates that move and interact with each other on top of the more molten **asthenosphere** in the mantle below. It is estimated that there are between 9 and 15 major plates and up to 40 minor plates. Some of these plates are **oceanic**, which means they contain an ocean basin, while others are **continental** and carry a landmass. The line formed by the meeting of two plates is called a **fault**. A well-known example is the San Andreas Fault, where the Pacific and North American plates meet. These faults are classified as **convergent** (plates colliding into each other) or **divergent** (plates moving away from each other). **Transform boundaries** occur when two plates slide past each other in opposite directions horizontally.

The larger plates are:

- North American plate
- Eurasian plate
- Pacific plate
- Australian plate
- Antarctic plate
- South American plate
- African plate
- Indian plate

Major plates of the lithosphere

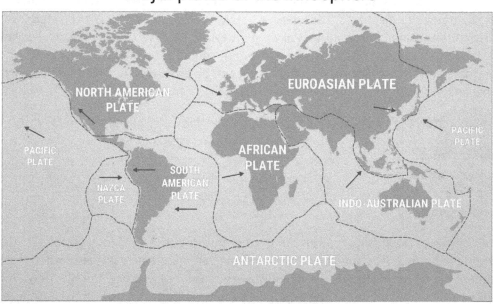

For the most part, these plates contain the countries/continents for which they are named. Scientists theorize that these plates were once part of a supercontinent, called **Pangea**, that existed over 175 million years ago. Evidence for Pangea's existence comes from geological findings, corresponding shapes of adjacent but currently independent landmasses, and fossils of identical species found in areas that used to be joined but have since separated.

Planet Earth is not a perfect sphere; it's actually slightly ellipsoid or oblong (like a football), and it's divided into three main layers from surface to center: the crust, mantle, and core. The crust and core are solid, while the mantle possesses a more fluid quality. In plate tectonics, the plates are considered to be the Earth's **lithosphere**, which is the crust and uppermost solid portion of mantle. One can visualize the plates as flat planks of wood covering the surface of a pool bobbing up and down and bumping, sliding, or moving away from each other. These different plate interactions create Earth's landscape.

At convergent boundaries, the collision of two plates can give rise to mountain ranges. Another possibility is that one plate slides atop another and pushes it down, creating magma and volcanoes, a process called **subduction**. This frequently happens when an oceanic and continental plate converge, and the denser oceanic plate sinks below under the continental plate. Volcanic activity also occurs at divergent boundaries—when the two plates separate—and magma escapes to the surface and solidifies.

This is how new land is formed. Faults are not the only places a volcano may appear; plates may contain areas called **hot spots**, and when these hot spots reach the surface, a volcano forms. Earthquakes are another consequence of plate tectonics, occurring as a result of sliding or colliding plates. Friction energy created between two plates can become an earthquake.

Geography

Geography is the study of the Earth's layout and features. There are imaginary lines that circle the globe and divide it into sections that help people track time and navigate. These lines run both vertically and horizontally from pole to pole. Vertical lines, also called **meridians**, measure east/west distance and are described as **longitudinal**. They are 15 degrees apart from each other, and the first line or point of zero, designated the Prime Meridian, serves as the reference line. Horizontal lines measure north/south distance and are described as **latitudinal**. The central latitude line is the **equator**—the line between the Northern and Southern Hemispheres. Each degree between these lines is further divided into sixty intervals called **minutes**, and each minute is divided into sixty intervals called **seconds**. For latitudinal lines, degree count increases as one moves away from the equator toward either pole. Using these lines and the coordinate system, any location on the planet can be precisely described; the advent of GPS technology has made this process much more user-friendly.

Aside from the equator, the Earth has four other latitudinal lines with specific names:

- Tropic of Cancer
- Tropic of Capricorn
- Arctic Circle
- Antarctic Circle

The Tropic of Cancer, or Northern Tropic, is the northernmost latitudinal line. It lies directly beneath the Sun during the June solstice (the specific date in June can vary each year). It is also called the northern solstice or summer solstice (in the Northern hemisphere) because it is when the Northern Hemisphere is most tilted toward the Sun and marks the beginning of summer for the Northern Hemisphere.

The **Tropic of Capricorn,** or Southern Tropic, is the southernmost latitudinal line. During the December or southern solstice, the Northern Hemisphere is tilted away from the Sun, while the Southern Hemisphere is tilted towards it. This date marks the beginning of winter for the Northern Hemisphere and summer for the Southern Hemisphere.

The **Arctic Circle** is the northernmost major circle of latitude. When the Northern Tropic receives light from directly overhead on the June solstice, the Sun doesn't fall below the horizon at or above the Arctic Circle, so that 24 hours of daylight are experienced.

The **Antarctic Circle** is the southern counterpart of the Arctic Circle. On the December solstice, people at, or below, the Antarctic Circle see that the Sun stays above the horizon for 24 hours. On the same date on the opposite pole (above the Arctic Circle), the Sun does not rise during the full day.

A good way to think about the major circles of latitude is remembering they are based upon how much direct sunlight (solar radiation) they receive at certain times of the year. Earth rotates on an axis that is tilted 23.5 degrees while it revolves around the Sun. This tilt causes the seasons experienced on Earth. Earth can be imagined as a spinning top that is teetering side to side; when the top leans toward the Sun, it gets more light, and the bottom is hidden. Then when it leans away from the Sun, the bottom

gets more light, and the top is hidden. Throughout this motion, the equator remains almost directly under the Sun all year and doesn't experience four different seasons. That's why areas close to the equator are warm all year round.

The Earth is also divided into different biomes or ecosystems. The simplest classification dictates that there are five different ecosystems:

- **Aquatic**: Contains marine and freshwater systems and found at all latitudes
- **Desert**: Areas of low precipitation and usually windy; can be found at nearly all latitudes
- **Forest**: Tropical, temperate, or boreal; types of trees depend on latitude
- **Grassland**: Prairies, savannahs, steppes; can exist in tropical, temperate, and cold climates
- **Tundra**: Found far from the equator toward either pole; receives the least amount of total annual sunlight

Weather, Atmosphere, and the Water Cycle

The study of the Earth's weather, atmosphere, and water cycle is called **meteorology**. The weather experienced in an area at a given time is a product of Earth's atmosphere above that location. The different atmospheric layers are defined by their temperature but are thought of by their distance above sea level (listed from lowest to highest):

- **Troposphere**: Sea level to 11 miles above sea level
- **Stratosphere**: 11 miles to 31 miles above sea level
- **Mesosphere**: 31 miles to 50 miles above sea level
- **Ionosphere**: 50 miles to 400 miles above sea level
- **Exosphere**: 400 miles to 800 miles above sea level

Above the exosphere is outer space. Together, the ionosphere and exosphere are considered the **thermosphere**. The ozone layer lies within the stratosphere, while the troposphere is where the conditions that lead to the observable weather on Earth originates. The manner in which the temperature changes is different in each of these layers. In the troposphere, it gets colder as the layer gets further from sea level, while the stratosphere gets warmer, the mesosphere gets colder, and the thermosphere gets warmer in this same direction.

The **atmosphere** is a layer of gas particles floating in space. The atmospheric levels are created by gravity and its pull on those particles. The Earth's atmosphere is mostly comprised of nitrogen and oxygen (78% and 21%, respectively) along with significantly lower amounts of other gases including 1% argon and 0.039% carbon dioxide. Gas particles have mass, and pressure is higher at the bottom (at the surface of Earth) because of all that mass in the layers above. Therefore, the farther away one gets from Earth's surface, the air gets thinner, as there are fewer air molecules compressing lower particles together. That's why breathing becomes more difficult when a person climbs to higher altitudes. To visualize this, one can imagine that when a person takes a breath, he or she is breathing in a cup of air. At higher altitudes, that cup is still the same size; it just holds fewer particles of gas.

Weather is a state of the atmosphere at a given place and time based on conditions such as air pressure, temperature, and moisture. Weather includes conditions like clouds, storms, temperature, tornadoes, hurricanes, and blizzards. In a given location on Earth, the weather experienced varies day-to-day, based on atmospheric conditions. The average weather for a particular area over a long period (usually over 30 years) is defined as that area's **climate**. The main force driving weather is the atmospheric variations between different areas on Earth. The major circles of latitude experience these differences because of the Sun. When there's a large difference in temperature between two areas, a jet stream forms. Pressure differences occur when there are temperature differences caused by surface variations (like mountains and valleys).

The **water cycle** is another factor that drives weather. It is the movement of water above, within, and on the surface of the Earth. During any phase of the cycle, water can exist in any of its three phases: liquid, ice, and vapor.

The processes that drive the cycle are:

- **Precipitation**: Rain, snow, hail, and sleet
- **Canopy interception**: Precipitation that falls on trees and doesn't hit the ground, eventually evaporating back into the atmosphere
- **Snowmelt**: Runoff from melting snow
- **Runoff**: Water moving across land that either seeps into the ground, evaporates, gets stored as lakes, or gets extracted by living organisms. It also includes surface and channel runoff.
- **Infiltration**: Water that moves from ground surface into the ground
- **Subsurface flow**: Moving water underground
- **Transpiration**: Release of vapor into the atmosphere from plants and soil
- **Percolation**: Water that flows down through soil and rocks
- **Evaporation**: Transformation from liquid water to gaseous vapor driven primarily by solar radiation
- **Sublimation**: Transformation of ice to gaseous vapor, never becoming a liquid
- **Deposition**: Transformation of vapor to solid ice
- **Condensation**: Transformation of vapor to liquid water

Water in the atmosphere exists as clouds, which are visible masses made of water droplets, tiny crystals of ice, dust, and various chemicals. The study of clouds, called **nephology**, is a subspecialty of meteorology. There are many types of clouds, and they can be classified based on the altitude at which they occur. It's important to note that clouds primarily occur in the troposphere. The classes of clouds are:

- **High-Clouds**: Occurring between 5,000 and 13,000 meters above sea level
    - **Cirrus**: Thin and wispy "mare's tail" appearance
    - **Cirrocumulus**: Rows of small puffs
    - **Cirrostratus**: Thin sheets that cover the sky
- **Middle clouds**: Occurring between 2,000 and 7,000 meters above sea level
    - **Altocumulus**: Gray and white and made up of water droplets
    - **Altostratus**: Grayish or bluish gray clouds
- **Low clouds**: Occurring below 2,000 meters above sea level
    - **Stratus**: Gray and cover the sky
    - **Stratocumulus**: Gray and lumpy low-lying clouds
    - **Nimbostratus**: Dark gray with uneven bases that occur with rain or snow

Here are some examples of different types of clouds:

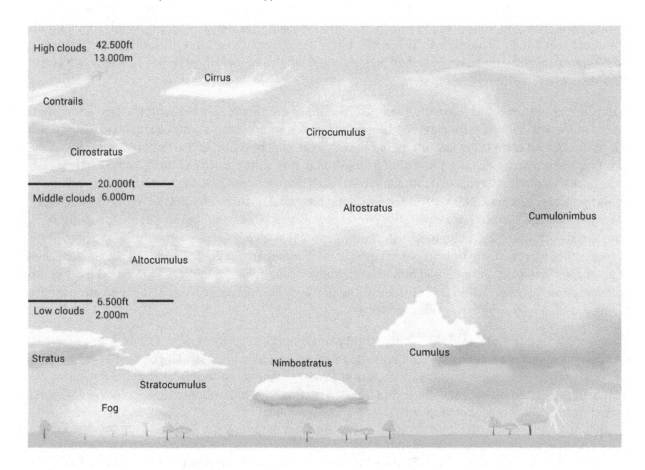

The vast majority of water on Earth is found in its oceans, which contain salt water, unlike lakes and other bodies, which contain freshwater. Salt water has a higher density than freshwater due to the dissolved salt. The total mass of all oceanic water is estimated to be 1.4×10^{24} grams, covering

approximately 70% of the Earth's surface (361,254,000 km²). The deepest part of the ocean is an area called the Challenger Deep in the Mariana Trench where the ocean floor is nearly 11,000 meters below sea level. The major oceans are the Pacific, Atlantic, and Indian Oceans. The Arctic and Southern Oceans are near each pole and are smaller.

Astronomy

Astronomy is the study of celestial bodies, or objects in space, and how they interact with each other. It is one of the oldest sciences; people have always looked up at the night sky with curiosity and amazement. Earth is a celestial body; others include the Sun, moon, other planets, black holes, satellites, asteroids, meteors, comets, stars, and galaxies. Information about celestial bodies and processes is obtained from observation via visible light and radiation throughout the universe. These observations are used in a field called **theoretical astronomy**, where scientists create theoretical models to explain or predict celestial processes or events.

From what astronomers have observed, the size of the universe is believed to be 91 billion light years. That means that if a person stood at the edge of the on a light at the opposite edge of the universe, it would take 91 billion years for the light to reach the person. And it's believed that the universe is constantly enlarging. A popular theory is that the universe was born 13 billion years ago, after an explosion called the **Big Bang**, and that the debris from that explosion (planets and stars) has been floating away from the epicenter ever since. It is also believed that the Universe is mainly made up of dark energy (about 73%) along with 23% dark matter and 4% regular matter, including stars, planets, and living organisms. The space between planets, stars, and galaxies is **interstellar space**, which is filled with **interstellar medium** (gas and dust).

Objects in the universe, including the Earth, are clumps of matter. Consistent with the laws of physics and gravity, these clumps form clusters. These clusters are solar systems, galaxies, galaxy clusters, superclusters, and something astronomers call the Great Wall of Galaxies. Earth's solar system is a planetary system on an ecliptic plane with a large sun in the center that provides gravitational pull. The Sun is primarily made of hydrogen and helium—metals comprise only 2% of its mass. It's 1.3 million kilometers wide, weighs 1.989×10^{30} kilograms, and has temperatures of 5,800 Kelvin (9980 degrees Fahrenheit) on the surface (also called the photosphere) and 15,600,000 Kelvin (28 million degrees Fahrenheit) in its core. It is the Sun's huge size and mass that give it so much gravity—enough to compress the hydrogen and helium (which exist as gas on Earth) into liquid form. The **Sun** is basically a series of giant explosions creating light and heat, and it's that huge gravitational force that pulls in those explosions and maintains the Sun's structure. The pressure at the Sun's core is 250 billion atmospheres, making it over 150 times denser than Earth's water!

Astronomers used to think there were nine total planets in the Solar System, but Pluto is now considered a dwarf planet, along with Ceres, Haumea, Makemake, and Eris. The eight planets can be divided into four inner (terrestrial) and four outer (Jovian) planets. Generally, the terrestrial planets are small, and the Jovian planets are larger and less dense with rings and moons. Listed in order of closest to the Sun to furthest, the planets are:

- **Mercury**: The smallest planet in the system and the one that is closest to the Sun. Because it's so close, it only takes about 88 days for Mercury to completely orbit the Sun It has a large iron core and the surface has craters like Earth's moon. There's no atmosphere, and it doesn't have any orbiting moons/satellites. From Earth, Mercury looks bright.

- **Venus:** The second planet is bright and has about the same size, composition, and gravity as Earth. It orbits the Sun every 225 days Its atmosphere creates clouds of sulfuric acid, and there's even thunder and lightning.

- **Earth:** The third planet orbits the Sun every year (about 365 days). Scientists believe it's the only planet in this system that's capable of supporting life.

- **Mars:** The Red Planet looks red because there's iron oxide on the surface. It also takes around 687 days to complete its orbit around the Sun. Interestingly, a day on Mars (one rotation about its axis), is very similar to the 24-hour day on Earth. Mars has a thin atmosphere as well as the largest mountain, canyon, and crater that astronomers have ever been able to see. Volcanoes, valleys, deserts, and polar ice caps like those on Earth have also been seen on its surface.

- **Jupiter:** The largest planet in the solar system is comprised mainly of hydrogen and helium (helium makes up 25% of its mass). The atmosphere on Jupiter has band-like clouds made of ammonia crystals that create tremendous storms and turbulence on the surface. Winds blow at around 100 meters per second, or over 220 miles per hour!

- **Saturn:** The second-largest planet is comprised mainly of hydrogen and helium along with other trace elements. The core is believed to be rock and ice. Saturn has a layer of metallic hydrogen. Winds are even stronger than those on Jupiter, reaching up to 1,100 miles per hour. Saturn has 61 moons, but the planet is most famous for its beautiful rings. There's no definitive explanation for how the rings formed, but two popular theories say they could be remnants from when Saturn itself formed, or they were moons that were destroyed in the past.

- **Uranus:** This planet has the coldest atmosphere of any in the Solar System, with a temperature that reaches -224.2 degrees Celsius (-371.56 °F). It's also mainly made of hydrogen and helium, but it has water, ammonia, methane, and even some hydrocarbons (the material humans are made of). A solid surface has yet to be observed through the thick layer of gas covering the planet. Uranus has 27 known moons.

- **Neptune:** The furthest planet is the third-largest by mass, and the second-coldest. It has 12 orbiting moons, an atmosphere like Uranus', a Great Dark Spot, and the strongest recorded winds of any planet in the system (reaching speeds of 2,100 kilometers per hour).

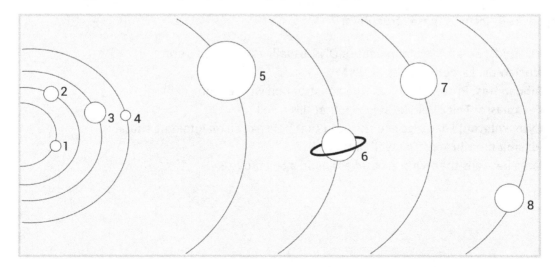

The nearest star to Earth's solar system is Proxima Centauri, which is about 270,000 Astronomical Units away. The closest object to Earth is the Moon, which is about 384,401 kilometers or 238,910 miles away. The Moon has two phases as it revolves around Earth—the waxing phase and the waning phase—which each last about two weeks. During the **waxing** phase, the Moon goes from new (black) to full moon; then it **wanes**, going from full to black. The Moon appears white because it's illuminated by the Sun. The edge of the shadow on the Moon is called the **lunar terminator**. The phases of waxing and waning in the Northern Hemisphere are:

- Waxing: The right side of the Moon is illuminated
    - **New moon** (dark): The Moon rises and sets with the Sun
    - **Crescent**: Tiny sliver illuminated on the right side
    - **First quarter**: The right half is illuminated; its phase is due south on the meridian
    - **Gibbous**: More than half is illuminated on the right side
    - **Full moon**: Rises at sunset and sets at sunrise
- Waning: The left side of the Moon is illuminated
    - **Gibbous**: More than half is illuminated on the left side
    - **Last quarter**: Half-illuminated on the left side; rises at midnight and sets at noon
    - **Crescent**: Tiny sliver illuminated on the left
    - **New moon** (dark): Rises and sets with the Sun again

Biology

Biology is the study of living organisms—it's the study of life. It encompasses many different areas of science, all of which fall under the category of biology or biological science, several of which are addressed in this section.

Every living thing on this planet is made of cells, either as a single cell (**unicellular**) or as a conglomerate of millions of cells (**multicellular**). In general, cells are quite similar, but they can be split into two categories: prokaryotic and eukaryotic. **Prokaryotic** cells are comparatively quite small and do not have a nucleus. **Eukaryotic** cells contain a nucleus as well as other membrane-bound organelles, which essentially have different functions within the cell and compartmentalize the cell's materials. It's generally believed that unicellular organisms can either be prokaryotic or eukaryotic, but multicellular organisms are always eukaryotic.

Prokaryotic cells include these structures:

- **Plasmids**: Small, circular pieces of DNA usually found in bacteria
- **Nucleoids**: Larger bundles of DNA
- **Ribosomes**: Protein builders (the construction workers of a cell)
- **Cytoplasm**: Thick gel-like solution that fills a cell
- **Cytoskeleton**: Filament-like skeleton that helps a cell maintain its shape
- **Plasma membrane**: The wall of the cell
- **Flagella**: Tails that whip around and help a cell move

Most of these structures are also found in eukaryotic cells and, depending on the type of eukaryotic cell, the following structures, called **organelles**, may also be present:

- **Nucleus**: Holds most a cell's DNA

- **Endoplasmic reticulum**: Tubular organelle attached to the nucleus and may have ribosomes attached to its walls. It functions like a big factory, producing proteins and lipids.

- **Golgi apparatus**: Similar to an endoplasmic reticulum, but not attached to a nucleus. Its role in a cell is to pack and transport materials in the cell.

- **Vacuoles**: Storage bubbles inside a cell

- **Mitochondria**: Small organelles that use oxidative energy to produce adenosine triphosphate (ATP), which is the energy molecule for the cell

- **Chloroplasts**: Organelles found in plants that produce ATP through the process of photosynthesis

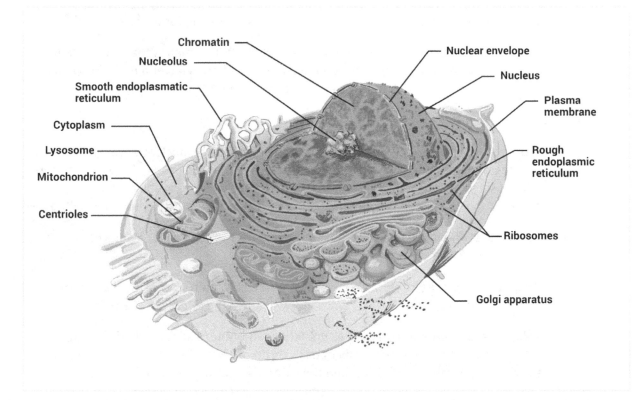

All cells contain genetic material in the form of **DNA** (deoxyribonucleic acid), and all DNA is made of the same material, although the specific arrangement of the DNA sequence can code for different genes and proteins. This is like how all books written in English have the same letters, but they can tell different stories. Prokaryotic cells hold their DNA in their cytoplasm in the form of a nucleoid. Eukaryotic cells contain most their DNA in a nucleus, and the rest can be found in either mitochondria or chloroplasts, depending on the type of cell. DNA exists inside the cell in the form of **chromosomes**, tightly wound bundles of DNA. Each species has a specific number of chromosomes that, all together, provide the blueprints that a cell requires to function, live, and reproduce. The number of chromosomes inside a cell

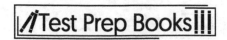

usually dictates how it reproduces. When a cell has only one set of chromosomes, it's considered to be **haploid** and reproduces through **binary fission**. If it has two sets of chromosomes (one from each parent), it's a **diploid** and goes through **mitosis**.

All cells reproduce by division: one cell splits and becomes two, then two to four, and so on. There are two methods of cell division: binary fission and mitosis. Prokaryotic cells can do either but usually divide via binary fission; eukaryotic cells only divide through mitosis. Binary fission is simple—as the cell is haploid, it just makes a copy of its DNA and splits in two. That's why bacteria can divide so rapidly. Mitosis is more complicated because eukaryotic cells are more complex than prokaryotes. Mitosis is divided into four different phases and, when combined with interphase and cytokinesis, forms the cell cycle (a cell's life):

- **Interphase**: The longest phase of the cell cycle, it occurs when the cell is living and functioning. Interphase is further divided into phases: G1, S, and G2

    o **G1**: First gap. The cell grows and functions.

    o **S**: Synthesis. The cell continues to grow, and it begins duplicating its DNA.

    o **G2**: Second gap. The cell is still growing and starts preparing for mitosis.

    o **G0**: Some cells (like neurons) enter a phase called G0. This means that the cell leaves the cell cycle and stops dividing. But, remembering that there's never a *never* in science, some cells can leave G0 and re-enter the cell cycle at G1.

- **Mitosis**: Cellular reproduction

    o **Prophase**: DNA begins to condense into chromosomes, the nuclear envelope starts to disintegrate, and two centrosomes start to move to opposite ends of the cell and form a mitotic spindle. Centrosomes are made of centrioles and microtubules, and they coordinate microtubule placement and use them to move things around inside the cell.

    o **Metaphase**: The mitotic spindle moves to the center of the cell, and pairs of chromosomes line up along the central spindle.

    o **Anaphase**: The pairs of chromosomes (sisters) start to pull apart. A cleavage furrow (groove) starts to form on the membrane/wall down the middle of the cell.

    o **Telophase**: The groove gets deeper until the membrane/wall is pinched, forming two cells containing one identical copy of each chromosome. The mitotic spindle disappears, the nuclear membrane reforms, and the chromosomes unpack to make chromatin.

- **Cytokinesis**: The actual splitting of the cell and all its contents to make two cells. It's not really a phase, but rather a process that starts during and finishes at the end mitosis.

Here's what the cycles look like:

The Cell Cycle

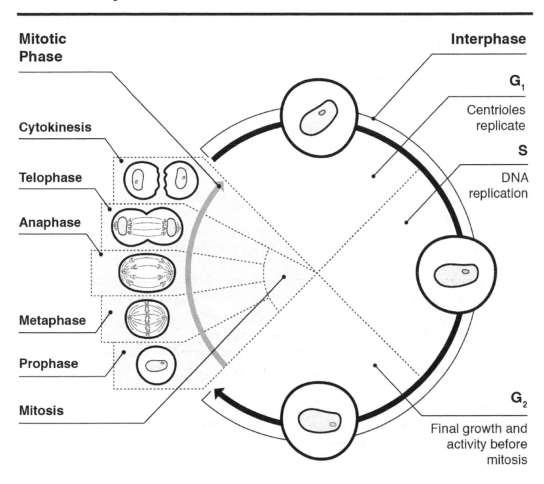

Cells of eukaryotic organisms used for sexual reproduction go through a process called meiosis. **Meiosis** only occurs in the **gametic** (sexually reproductive) cells and results in the production of sperm and oocytes (eggs). Meiosis is divided into two phases: Meiosis I and Meiosis II. Here is a timeline of meiosis in the cell cycle:

- **Interphase**: Like all other cells, contains the S-phase in which a cell duplicates its DNA.

- **Meiosis I**: Deals with homologous chromosomes (a pair containing a chromosome from each mother and father). It's further divided into phases:

  - **Prophase I**: The longest phase of meiosis

    - **Leptotene**: Chromosomes start to look like the letter "X" and consist of two sister chromatids (a chromatid is one side of the X).

    - **Zygotene**: Homologous chromosomes line up.

- **Pachytene**: Chromosome crossover begins. This process accounts for genetic recombination, wherein fragments of the mother's and father's genes are exchanged and allows an offspring to be a thorough mix of its parents, which increases genetic diversity.

- **Diplotene**: Homologous chromosomes start to separate from each other.

- **Diakinesis**: The chromosomes condense further.

- **Synchronous processes**: The two centrosomes appear and migrate to opposite sides of the cell.

- **Metaphase I**: Paired homologous chromosomes line up in the middle of the cell.
- **Anaphase I**: The pairs start to separate from one another.
- **Telophase I**: The completely separated pairs decondense, and the cell splits into two haploid cells.

- **Interkinesis or interphase II**: Rest period

- **Meiosis II**: Equational segregation, which is the splitting of sister chromatids

  - **Prophase II**: The nuclear envelope disappears, chromatids condense, and centrosomes appear.
  - **Metaphase II**: Sister chromatids line up in the middle of the cell.
  - **Anaphase II**: Sister chromatids start to separate.
  - **Telophase II**: Chromosomes are completely separated, they decondense, and the cells split (resulting in a total of four haploid cells).

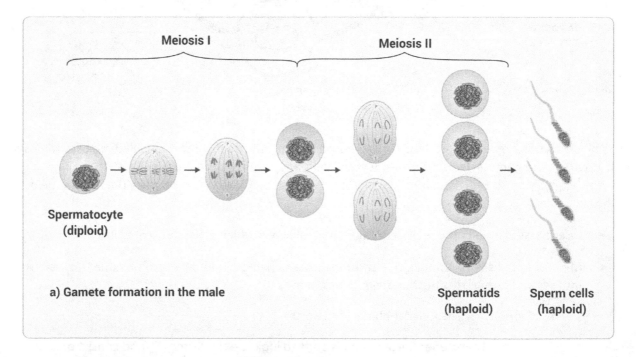

a) Gamete formation in the male

The two main classes of eukaryotes are plants and animals. Although a human is noticeably different from a carrot, cells in these organisms have only a few differences. The main differences are the cell structure, appearance, and method of energy production.

Plant cells have a **cell wall** made of cellulose—a strong, fibrous material that adds bulk to stool because it's so hard to break down. This material must be strong in order to prevent the cell from bursting when water enters and increases the intracellular pressure. Animal cells don't have a cell wall. Instead, they have a **cell membrane** composed of several materials (which differ depending on cell type) embedded in a phospholipid bilayer. This bilayer is exactly what it sounds like: two layers of phospholipids stacked tail-to-tail. A **phospholipid** is a molecule with a glycerol (phosphate) head and two fatty acid (lipid) tails. This membrane is much weaker compared a plant's cellulose wall, so animal cells are much more susceptible to bursting from water pressure. And herein lies an important characteristic of the phospholipid bilayer: the tails inside the layer are actually **hydrophobic** (water-repellent) so they are able to prevent too much water from entering and killing the cell.

The shapes and sizes of plant and animal cells are also quite different. Plant cells have a consistent rectangular shape and are generally much larger, ranging from 10–100 micrometers. Animal cells vary in shape and are around 10–30 micrometers. Human liver cells (which are round) look completely different from nerve cells (which have a large cell body and long projections outward). The vacuoles are also different; plant cells tend to have just one large vacuole, while animal cells have several tiny ones.

The most important difference between plants and animals is how they create energy. Animals must consume energy from other living organisms. Plants have the amazing ability to use energy from the sun to power their growth and survival. The little organelles that make this possible are the **chloroplasts** in plants and **mitochondria** in animals. Both use carbon, oxygen, and water. Through photosynthesis, a chloroplast is able to extract carbon from the carbon dioxide (CO_2) in the air and combine it with water (H_2O) to make glucose ($C_6H_{12}O_6$) and oxygen (O_2). Glucose provides energy for animal cells. In the body, mitochondria use the energy embedded inside glucose molecules and combine them with oxygen to create carbon dioxide, water, and ATP (a form of useable energy for physiological processes). This process is called cellular respiration. Memorizing the chemical equations for these reactions may help with understanding these processes:

- **Photosynthesis**: $6\ H_2O + 6\ CO_2 + sunlight \rightarrow C_6H_{12}O_6 + 6\ O_2$
  - Six molecules of water and six of carbon dioxide make one glucose molecule and six molecules of oxygen.
- **Cellular respiration**: $C_6H_{12}O_6 + 6\ O_2 \rightarrow 6\ H_2O + 6\ CO_2 + heat$ (energy)
  - Note that this is basically the reverse of photosynthesis.

Bacterial cells are a completely different story. There are a huge variety of bacteria, and they can actually share characteristics with both plant and animal cells. Some can utilize photosynthesis, some can only perform cellular respiration, and some can even do both or neither!

Genetics

Genetics is the study of how living organisms pass down traits to their offspring and future generations. It all starts with **DNA**—the universal language of genetic information. An organism's physical features and cellular instructions are written as **genes**, which are single units of genetic information that are stored in every cell as a set of chromosomes.

The discovery of DNA is usually credited to two scientists, Francis Crick and James Watson. However, before them, in 1869 at the University of Tübingen, Friedrich Miescher became the first person to ever identify and isolate DNA from a cell; at that time, he called it **nuclein**. In 1952, a doctoral student named Raymond Gosling was able to take an x-ray diffraction image of DNA known as **photograph 51**, famous for inspiring Watson and Crick. From that photograph, they were able to deduce the information needed to develop their model of DNA. In 1953, they correctly postulated that DNA has a double-helical structure and that it carried the genetic information. Watson and Crick won the Nobel Prize in Physiology or Medicine in 1962.

A **double helix** looks like a twisted ladder; it is twisted to save space. When looking at DNA, the side rails of the ladder are considered the backbones, and the rungs are pairs of nitrogenous bases. Each backbone is a string of sugars and phosphates covalently bonded together. By contrast, the bonds between paired bases is quite weak—they are attached via hydrogen bonds. If DNA is broken down into its building blocks, a pile of nucleotides is acquired. A single **nucleotide** contains one five-carbon sugar (pentose) and one phosphate group (backbone components) attached to a nitrogenous base (one half of a rung). So, if all the pieces of DNA look so similar, how can it vary enough to provide all the information needed to create a living organism?

The answer comes from the **nitrogenous bases**; the order in which the bases are placed in a DNA strand provides specific information. Here's the unbelievable part: There are only four different bases for DNA. That means a person's entire genetic code, along with every other single organism that has ever lived, is written with just four letters! It's important to note that human DNA is a combination of about 3 billion bases, and different species can have different DNA lengths. The two types of nitrogenous bases—the **purines** and the **pyrimidines** (along with an easy mnemonic to memorize which base belongs to each group) —are as follows:

- **Purine** – PURe *As* *G*old
  - Adenine (A)
  - Guanine (G)
- **Pyrimidines** – **Sharp PYRamids *CUT***
  - Cytosine (C)
  - Thymine (T)
  - Uracil (U) doesn't contribute to DNA—it's the fifth base found only in RNA

When double-stranded DNA forms, the unique order of bases is preserved because each purine can only form a hydrogen bond with a specific pyrimidine. When looking at the paired bases in each rung of the ladder, the only two pairs are A – T and G – C.

Here's a simplified example of a short segment of double-stranded DNA (dsDNA):

5' – ATCGTTTGAGCACTAGCG –3' (strand #1)

3' – TAGCAAACTCGTGATCGC –5' (strand #2)

Note how As are always present across Ts and Gs across Cs. Before DNA gets replicated, the two strands have to separate, creating this:

5' – ATCGTTTGAGCACTAGCG –3' (strand #1) + 3' – TAGCAAACTCGTGATCGC –5' (strand #2)

Replication involves grabbing unattached nucleotide bases and pairing them appropriately, creating this:

5' – ATCGTTTGAGCTCTAGCG –3' (strand #1)   3' – TAGCAAACTCGTGATCGC –5' (strand #2)

3' – TAGCAAACTCGAGATCGC –5' (new strand) **+** 5' – ATCGTTTGAGCACTAGCG –3' (new strand)

The two products are completely identical to the original (the one on the right just needs to be flipped so that the new strand is on top). The designations 5' and 3' provide a sense of direction when dealing with DNA. In organic chemistry, the nucleotide sugar ring's carbon atoms are numbered. When looking at strand #1 above, 5' on the left and 3' on the right means the sugars of that strand are connected with the 5th carbons on the left and the 3rd carbons are on the right. Moving toward the 5' end is called upstream and going toward the 3' is downstream. All of this is important because when cells replicate DNA, the new strand is created in a 5' to 3' direction (the 5' carbon has a phosphate group that provides the energy to form a covalent bond with the hydroxyl group on the 3' carbon). The 5' end is considered the "head" and the 3' end is the "tail" of the strand.

Remember:

- **Upstream**: Move from 3' to 5' (going up in #).
- **Downstream**: Move from 5' to 3' (going down in #).
- **Replication**: Read upstream, create downstream.

The collection of genes within an individual organism is called the **genotype**. Not all genes are expressed (recessive genes can be overpowered by dominant ones), and the physical manifestation of those genes is called a **phenotype**. The variations of genes are called **alleles**. The specific location of a gene sequence is referred to as a gene's **locus**.

An easy example for these terms is eye color. Everybody possesses a gene that dictates eye color. Each color is a different allele. If a boy received one blue eye allele and one brown eye allele from his mother and father, and the boy has brown eyes, that means he is phenotypically brown-eyed. It doesn't mean he doesn't have the allele for blue eyes; the blue-eye allele just isn't physically manifested. In this fictitious example, the gene for brown eyes is dominant while that for blue eyes is recessive, so brown eyes get expressed.

Humans have 23 pairs of chromosomes, one set from a father and one from a mother, giving each person pairs of the same allele or gene trait. That's called Mendel's first law of segregation. When studying genetics, an allele is represented by uppercase or lowercase letter depending on which trait is being studied. An uppercase letter means that allele is dominant, and a lowercase letter represents a recessive trait. In the eye color example, "B" represents brown eyes and "b" is blue eyes (because it's recessive). If, for simplicity, these are the only two alleles available, then the only possible pairs are BB, Bb, and bb. When a brown-eyed father (he happens to be Bb) and a blue-eyed mother (as blue eyes are recessive, she must be bb) have a child, they each give one of their two alleles. Mendel's second law of independent assortment dictates that each parent has a 50% chance of passing down either allele. The only exception to this law is if two genes are linked, which means that they are inherited together and don't follow proper segregation rules.

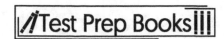

A Punnett square can be used to illustrate the possible combinations:

	Mom b	Mom b
Dad B	Bb	Bb
Dad b	bb	bb

The top row has the mother's alleles and the left column has the father's. The shaded squares represent the possible combinations offspring can be born with. If the child ended up getting a "b" from both parents, giving it bb, it is a **homozygous** pair. In this case, it would be recessive (blue eyes). If the child received Bb, its genotype would be considered **heterozygous**, and the child would have brown eyes because the brown-eyed gene (B) would dominate the blue-eyed gene (b).

Organism Classification

There are many different species that have existed and currently live on Earth, and scientists are still discovering more. A comprehensive database called the "Catalogue of Life" lists over 1.6 million species. To keep track of them all, a classification system was developed. The science of this system is called **taxonomy**, and it's based on an organism's characteristics.

The classes are (from broadest to most specific):

- **Domains**
  - **Bacteria**: Prokaryotic cells with membranes made of fatty acids with ester linkages
  - **Archaea**: Prokaryotic cells with some eukaryotic traits
  - **Eukarya**: Eukaryotic organisms (plants, animals, fungi, etc.)
- **Kingdoms**
  - **Monera**: All prokaryotic organisms. This kingdom actually covers both bacteria and archaea domains because the three-domain system was created more recently than the kingdom classification system.
  - **Protista**: Single-celled eukaryotic organisms
  - **Fungi**
  - **Plantae**
  - **Animalia**: Includes all animals and insects
- **Phylum/divisions**: Generally divided based on special body parts, such as phylum Chordata (having a nerve cord) and phylum Echinodermata (having spiny skin)

- **Classes**: Mammals are an example of a class
- **Orders**
- **Families**
- **Genus**: At this level, all organisms in a genus are fairly similar. An example is Lasioglossum, one of the many genera for bees. Usually, it is difficult for an average person to distinguish differences between species in a single genus. The name of a species' genus is the capitalized letter in its scientific name, such as F. catus (a house cat), which belongs to the genus Felis.
- **Species**

The science of taxonomy is important for the study of biodiversity and wildlife conservation. While there are no concrete rules for how new species are named, there are two sets of codes that regulate the process. They are the International Code of Zoological Nomenclature or ICZN (for animals) and the International Code of Nomenclature or ICN (for algae, fungi, and plants).

This classification system does not cover viruses because they are a form of non-cellular life. Viruses are unique because they can be considered both living and non-living. They contain DNA or other genetic material like a living organism, but they are unable to replicate by themselves. When a virus is outside of a host, it is just an object; once that virus is inside a host cell, it acquires the ability to replicate, modify, infect, and spread. Some viruses degrade or die when outside of a host, and some can exist indefinitely.

Anatomy

Anatomy is the science of the structure and parts of organisms. This guide focuses on human anatomy. Although people have studied the human body for hundreds of years, doctors and scientists are still trying to answer many questions. How exactly does the brain work? Why are cancer cells more likely to appear in specific populations? What makes a cell decide to be a liver cell instead of a neuron during development?

While the human body and a car look nothing alike, they are actually somewhat similar. A person is basically a machine. They both have systems or components that, when put together, can consume fuel and perform work or movement. The human body can be divided into ten different systems:

Cardiovascular/Circulatory System
The **cardiovascular/circulatory system** consists of the heart, blood (which carries nutrients and waste), and blood vessels. The heart's job is to pump oxygenated blood to the rest of the body through blood vessels called arteries. **Exchange** (delivering nutrients and extracting waste) at the tissues occurs in capillaries, which are the smallest blood vessels. When the body has used the oxygenated blood, it returns to the heart via vessels called veins. The heart can then circulate it through the lungs to be re-oxygenated.

There are four special chambers of the heart, and each has its own role. They are:

- **Right atrium**: Collects de-oxygenated blood returning from the body and holds it until the right ventricle is ready.

- **Right ventricle**: Receives blood from the right atrium and pumps it into the lungs to re-oxygenate it.

- **Left atrium**: Collects oxygenated blood from the lungs and holds it until the left ventricle is ready.

- **Left ventricle**: The most powerful chamber of the heart, the left ventricle receives blood from the left atrium and pumps it around the body. It must be strong in order to achieve the high pressure needed to circulate blood throughout the body.

Along with oxygen, blood (which consists of plasma, red blood cells, white blood cells, and platelets) carries nutrients and waste products through the body. The average adult has around five liters of blood in the body and, in a healthy individual at rest, that volume can be circulated through the body every minute! There are two special systems within the circulatory system called the hepatic and hypophyseal portal systems. They are special in that they are veins that carry deoxygenated blood to an organ other than the heart. The **hepatic portal** carries highly nutritional blood from the digestive system to the liver for processing before returning to the heart. The **hypophyseal portal** carries specific hormones between the hypothalamus and pituitary within the brain, so that these hormones don't need to travel through the rest of the body before reaching their destination.

Digestive System

The **digestive system** manages food consumption and processing, from chewing to nutrient absorption and defecation. Although the digestive system is inside the body, it is considered to be an open system because it's technically exposed to the outside (an uncut string can be seen from mouth to anus if a person swallowed it). This system consists of the mouth, throat, esophagus, stomach, small and large intestines, and anus. The mouth roughly chews or breaks down food with the teeth, tongue, and saliva so that it can move to the stomach via the esophagus. The stomach's job is to further break down the food so the intestines can absorb it. The small intestines are (from stomach to large intestine) the duodenum, jejunum, and ileum. Most nutrients are absorbed in the small intestine. The large intestine—consisting of the colon, rectum, and anus—absorbs any remaining nutrients and eliminates waste.

Endocrine System

The **endocrine system** is the hormonal or glandular system of the body. Organ systems communicate with each another via hormones that travel in the bloodstream. This system is vital for maintaining homeostasis (equilibrium of the body). Like the circulatory system, it is intimately involved with all other systems of the body. The endocrine system consists of the hypothalamus, pancreas, ovaries, testes, gastrointestinal tract, and the adrenal, parathyroid, pineal, pituitary, and thyroid glands. These organs are all endocrine glands (they secrete into the blood), which differ from exocrine glands, which secrete outside the body (such as sweat and salivary glands). An example of a vital hormone is epinephrine or adrenaline, which is excreted by the adrenal glands atop the kidneys. When people feel danger, increased adrenaline can aid chances of survival by making them alert and increasing blood flow to muscles for quick actions.

Integumentary System

The **integumentary system** is the skin and all things contained within it (hair, nails, sensory receptors, and sweat and oil glands). The skin is the largest organ of the human body, and its main function is to protect everything inside, much like the chassis of a car protects its passengers. Skin also cushions the body, excretes waste, and regulates body temperature. It is composed of the epidermis (the outer layer), dermis, and hypodermis. The epidermis is composed of cells called keratinocytes (making it waterproof) and melanocytes (the cells that give a person a tan or dark color). At the base of the epidermis is a layer of basal cells, which replicate to form the epidermis. The dermis, or middle layer, is composed of connective tissue and holds the sweat and oil glands, touch and temperature sensors, and hair follicles. The hypodermis, also known as the subcutaneous layer, provides cushioning and contains fat, nerves, and blood vessels.

Immune/Lymphatic System

The **immune** and **lymphatic system** work together to fight off and clean out elements that are potentially harmful to the body; they consist of white blood cells, lymph nodes, lymphatic vessels, and select organs such as the spleen. Lymphatic vessels are similar to veins, but instead of carrying blood, they carry waste fluid from the body. Along these vessels are lymph nodes that hold white blood cells, which protect the body from foreign invaders. There are five types of white blood cells: monocytes, neutrophils, lymphocytes (includes B-cells and T-cells), eosinophils, and basophils. Monocytes and neutrophils eat and destroy invaders like bacteria and viruses. Lymphocytes (B-cells) produce antibodies to protect the body from infections.

Musculoskeletal System

The **musculoskeletal system** contains the muscles, skeleton, cartilage, tendons, and ligaments. In addition to enabling movement, it protects vital organs, such as the heart and lungs inside the ribcage and the brain inside the skull. An adult skeleton has 206 bones, which are classified as long, short, flat, irregular, or sesamoid (small and sesame seed-like) bones. Bones are connected at joints like the elbow or knee. The connective tissue joining bone to bone is called a ligament. A tendon is connective tissue that links muscle to bone. Cartilage is a material that lines a joint to provide cushioning and reduces friction during movement; it also forms structural components of certain tissues like the eyes. Other vital roles of the skeletal system are storing calcium for the body and holding marrow, where red and white blood cells are produced.

Nervous System

The **nervous system** is the body's brain and nerves, similar to a modern car's onboard computer. It's split into two parts: the **central nervous system** (CNS) and **peripheral nervous system** (PNS). The CNS consists of the brain and spinal cord. The brain allows for thought, memory, and emotion. It also provides the electrical impulses that tell the rest of the body what to do. It sends and receives these electrical signals to the body via the PNS. The nervous system is also divided into the **somatic** (voluntary) and **autonomic** (involuntary) divisions based on function. The somatic nervous system contains the nerves and parts of the brain that control voluntary movement. The autonomic system consists of the sympathetic and parasympathetic nervous systems, the "fight or flight" and "rest and digest" systems, respectively.

Renal/Urinary System

The **renal/urinary system** consists of the kidneys, ureters, bladder, and urethra. The kidneys filter excess water and waste—such as ammonia—from the blood to make urine, which travels to the bladder through a tube called the ureter. From the bladder, urine is excreted from the body via the urethra. Although people are normally born with two kidneys, it is possible to survive with just one.

Reproductive System

Consisting of the sex organs, the **reproductive system** is the only system that differs in males and females. Males have a penis, prostate, and scrotum containing the testes. Females have a vagina, cervix, uterus, ovaries, fallopian tubes, and mammary glands (in the breasts). Aside from producing and protecting sperm and oocytes (eggs) for reproduction, testes and ovaries make the hormones (testosterone and estrogen) essential for the development of male and female secondary sexual characteristics, such as the Adam's apple and breasts. Sex cells replicate via meiosis, creating haploid cells. When a sperm cell fertilizes an oocyte, the two haploid sets of chromosomes create a diploid set, which develops into a baby.

Pulmonary/Respiratory System

Comprised of the nose, mouth, throat, trachea, lungs, and diaphragm, the **pulmonary/respiratory system** facilitates the exchange of gases—mainly carbon dioxide and oxygen—in a process called **respiration**. Oxygen-rich air enters the nose and mouth and travels down the trachea to the lungs. Structures within the lungs called alveoli (air sacs) allow red blood cells to exchange the carbon dioxide produced by the body with oxygen through diffusion. By themselves, the lungs are unable to actively expand or contract to pull in and expel air (ventilate). The actual work is done by the diaphragm, a layer of muscle that makes up the floor of the chest cavity that holds the heart and lungs. When a person inhales, the diaphragm contracts and straightens, which creates negative pressure inside the pleural cavity and pulls the lungs open to draw in air from the environment. During exhalation, the diaphragm relaxes and moves up into the chest cavity, which releases the negative pleural pressure and allows the lungs to expel air. Breathing is an involuntary process, meaning that it is controlled by the autonomic nervous system—specifically an area of the brain called the brain stem.

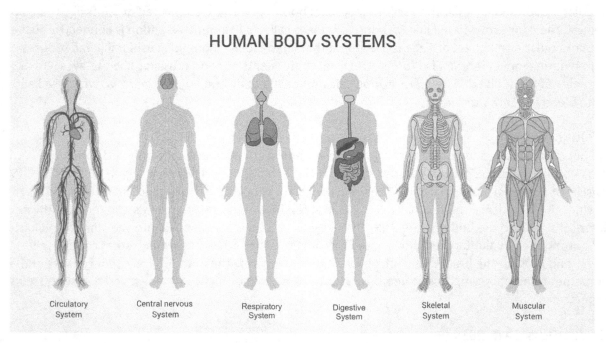

HUMAN BODY SYSTEMS

Circulatory System   Central nervous System   Respiratory System   Digestive System   Skeletal System   Muscular System

Biological Relationships

The study of biological relationships is called **ecology**, which explores how organisms affect one another. These relationships can be defined as either **intraspecific** (within the same species) or **interspecific** (between different species) interactions. For example, a man being friendly to his neighbor can be considered an intraspecific interaction, while a man interacting with a dog is an interspecific interaction.

These can also be categorized in terms of the effects they have on participating organisms:

- **Competition**: This is the basis for Darwin's theory of natural selection and survival of the fittest. Competition occurs when the organisms' interactions are mutually harmful, as they compete with one another in order to benefit themselves.

- **Amensalism**: Occurs when an organism harms another, but the organism inflicting harm does not benefit from the interaction.

- **Antagonism**: One individual or species hunts another (inflicting harm) for the benefit of food. The harmed species is not working toward the same goal, differentiating amensalism from competition. Classic examples of antagonism are predation and parasitism.

- **Neutralism**: Occurs when different individuals or species interact with one another and neither species is harmed or benefits from the relationship.

- **Facilitation**: Describes interactions in which one participant benefits without harming the other.

  o **Commensalism**: Occurs when only one participant receives benefit from the interaction.

  o **Mutualism**: Occurs when both parties benefit from the interaction.

- **Symbiosis**: A close mutualistic relationship between organisms of different species. A popular example is the relationship between a clown fish and a sea anemone. The clown fish cleans the sea anemone, and in return, it is protected from predators by the anemone's stinging arms.

Another large area of ecological studies is the **food chain**—the network of links between organisms. Producer organisms like grass or algae make up the bottom of the chain, while the top is comprised of apex predators such as lions and tigers. The food chain is useful for depicting the flow of energy, or nutrition, from bottom to top. Terms commonly used while discussing the food chain include:

- **Autotroph**: A producer in the food chain with the ability to produce organic compounds from the environment. Photosynthetic plants are examples of autotrophs.

- **Heterotroph**: A consumer in the food chain that must consume organic materials to thrive. Humans are examples of heterotrophs.

- **Lithotroph**: An organism with the ability to use inorganic material to survive. Lithotrophs are exclusively microbes, such as those that survive deep underwater, where they utilize inorganic sulfur that escapes from volcanic vents.

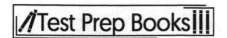

Biological Concepts

Biological concepts include all the terms discussed in earlier sections concerning biology, as well as:

- **Territoriality**: The behavior of an organism that claims a specific sociographical area/location and defends it, usually because its food or sexual mates are in that area.

- **Community**: A collection of individual organisms living and interacting within a specific area.

- **Niche**: The role of a species in a community.

- **Dominance**: Refers to the most prevalent species in a community.

- **Altruism**: A biological interaction in which one individual or species provides benefit to another at a cost to itself. A wealthy man is considered altruistic when he donates money to the poor.

- **Threat display**: Behavior of an organism with the intent to scare other individuals.

- **Competitive exclusion**: Also called Gause's Law, which dictates that when two species within the same community are competing for the same resource, they cannot stably coexist. In other words, one will dominate the other.

- **Species diversity**: Refers to the number of different species in a community.

- **Biotic**: Living organisms in a community, such as other species.

- **Abiotic**: Non-living factors such as terrain and weather.

- **Ecosystem**: Encompasses a community and all its environmental factors.

- **Biomass**: Mass of organic matter within an ecosystem.

- **Mimicry**: An organism that adapts the appearance or actions of a different species in order to avoid predators or harm.

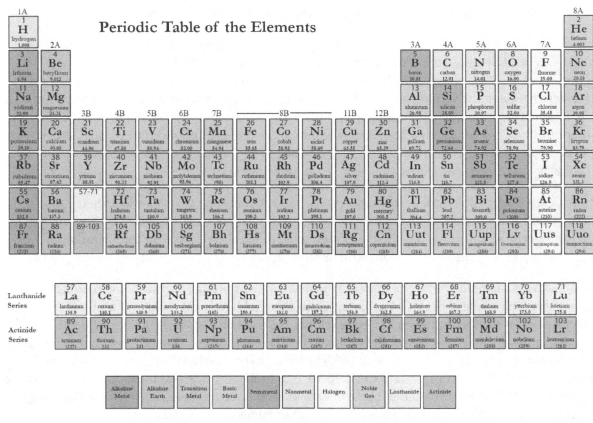

Chemistry

Periodic Table

The **periodic table** is a chart of all 118 known elements. The elements are organized according to their quantity of protons (also known as their atomic number), their electron configurations, and their chemical properties. The rows are called periods, and the columns are called groups. Elements in a given group have similar chemical behaviors. For example, Group 8A is the noble gases, and because their outer electron shell is full, they are non-reactive. The closer the element is to having a full set of valence electrons, the more reactive it is. Group 1, the alkali metals, and Group 7, the halogens, are both highly reactive for that reason. The alkali metals form **cations** and lose their lone electron while halogens pick up an electron and become **anions**.

In each box in the period table, an element's symbol is the abbreviation in the center, and its full name is located directly below that. The number in the top left corner is the atomic number, and the atomic mass of the element is the number underneath the name. The atomic mass of the element is noted in atomic mass units, or amu, which represents the number of protons and neutrons combined. The number of protons defines the element, but the mass number can be different due to the existence of elements with a varying number of neutrons, which are called isotopes. The amu shown on the periodic table is a weighted mass (based on abundance) of all known isotopes of a particular element.

Atomic Structure

Atoms are made up of three subatomic particles: protons, neutrons, and electrons. The **protons** have a positive charge and are located in the nucleus of the atom. **Neutrons** have a neutral charge and are also located in the nucleus. **Electrons** have a negative charge, are the smallest of the three particles, and are located in orbitals that surround the nucleus in a cloud. An atom is neutral if it has an equal number of electrons and protons. If an atom does not have an equal number of electrons and protons, it is an **ion**. When there are fewer electrons than protons, leaving it with a positive charge, it is termed a **cation**. When there are more electrons than protons, leaving a negative charge, it is an **anion**.

There are many levels of orbitals that surround the nucleus and house electrons. Each orbital has a distinct shape and three quantum numbers that characterize it: n, l, and m_l. The n is the principal quantum number and is always a positive integer, or whole number. As n increases, the size of the orbital increases, and the electrons are less tightly bound to the nucleus because they are further away. The angular momentum quantum number is represented by l and defines the shape of the orbital. There are four numerical values of l that correspond to four letter representations of the orbital: 0 or s, 1 or p, 2 or d, and 3 or f. Lastly, m_l represents the magnetic quantum number and describes the orientation of the orbital in space. It has a value in between -1 and 1, including zero.

Electrons exist outside the nucleus in energy levels, and with each increasing period, there is an additional energy level. Hydrogen and helium are the only elements that have only one energy level, which is an s orbital, which can only hold two electrons. Their relative electron configurations are $1s^1$ and $1s^2$. Each orbital holds two electrons with opposite spins.

Energy level two can hold a total of eight electrons: two in the s orbital, and six in the p orbital, of which there are three, since electrons pair up: $p_x p_y p_z$. All alkali metals and alkali Earth metals have an electron configuration that ends in either s^1 or s^2. For example, magnesium ends in $3s^s$ and cesium ends in $6s^1$. Groups 3A to 8A all end in the Xp^y configuration, with X representing the energy level and y representing how for inland it is. For example, oxygen is $2p^4$. The third energy level has s, p, and d orbitals, and together can hold 18 electrons, since d can hold 10. D orbitals start with the transition metals, and even though it appears that they are in the fourth energy level, they actually are on the third. The energy level for the whole d block is actually the period number minus one. For example:

Mn: $1s^2 2s^2 2p^6 3s^2 3p^6 4s^2 3d^5$

The f block, which can hold 14 electrons, has a similar pattern for its period number, except two is subtracted from the period number instead of one.

Acids and Bases

Acids and bases are defined in many different ways. An **acid** can be described as a substance that increases the concentration of H^+ ions when it is dissolved in water, as a proton donor in a chemical equation, or as an electron-pair acceptor. A **base** can be a substance that increases the concentration of OH^- ions when it is dissolved in water, accepts a proton in a chemical reaction, or is an electron-pair donor.

Autoionization of Water

Water can act as either an acid or a base. When mixed with an acid, water can accept a proton and become an H_3O^+ ion. When mixed with a base, water can donate a proton and become an OH^- ion. Sometimes water molecules donate and accept protons from each other; this process is called **autoionization**. The chemical equation is written as follows: $H_2O + H_2O \rightarrow OH^- + H_3O^+$.

PH Scale

The **pH scale** is a numeric scale that determines whether a solution is acidic, basic, or neutral. The pH of a solution is equal to the inverse base 10 logarithm of its activity of H^+ ions ($pH = \log_{10}(1/(H^+ \text{ activity}))$). Solutions with a pH value greater than seven are considered basic and those with a pH value less than seven are considered acidic. Pure water is neutral and has a pH of 7.0.

Strength of Acids and Bases

Acids and bases are characterized as strong, weak, or somewhere in between. Strong acids and bases completely or almost completely ionize in an aqueous solution. The chemical reaction is driven completely forward, to the right side of the equation, where the acidic or basic ions are formed. Weak acids and bases do not completely disassociate in an aqueous solution. They only partially ionize and the solution becomes a mixture of the acid or base, water, and the acidic or basic ions. Strong acids are complemented by weak bases, and vice versa. A conjugate acid is an ion that forms when its base pair gains a proton. For example, the conjugate acid NH_4^+ is formed from the base NH_3. The conjugate base that pairs with an acid is the ion that is formed when an acid loses a proton. NO_2^- is the conjugate base of the acid HNO_2.

Chemical Reactions

Types of Chemical Reactions

Chemical reactions are characterized by a chemical change in which the starting substances, or **reactants**, differ from the substances formed, or **products**. Chemical reactions may involve a change in color, the production of gas, the formation of a precipitate, or changes in heat content. The following are the five basic types of chemical reactions:

- **Decomposition Reactions**: A compound is broken down into smaller elements. For example, $2H_2O \rightarrow 2H_2 + O_2$. This is read as, "2 molecules of water decompose into 2 molecules of hydrogen and 1 molecule of oxygen."

- **Synthesis Reactions**: Two or more elements or compounds are joined together. For example, $2H_2 + O_2 \rightarrow 2H_2O$. This is read as, "2 molecules of hydrogen react with 1 molecule of oxygen to produce 2 molecules of water."

- **Single Displacement Reactions**: A single element or ion takes the place of another element in a compound. It is also known as a substitution reaction. For example, $Zn + 2 HCl \rightarrow ZnCl_2 + H_2$. This is read as, "zinc reacts with 2 molecules of hydrochloric acid to produce one molecule of zinc chloride and one molecule of hydrogen." In other words, zinc replaces the hydrogen in hydrochloric acid.

- **Double Displacement Reactions**: Two elements or ions exchange a single element to form two different compounds, resulting in different combinations of cations and anions in the final compounds. It is also known as a metathesis reaction. For example, $H_2SO_4 + 2 NaOH \rightarrow Na_2SO_4 + 2 H_2O$

  - Special types of double displacement reactions include:

    - **Oxidation-Reduction (or Redox) Reactions**: Elements undergo a change in oxidation number. For example, $2 S_2O_3^{2-} (aq) + I_2 (aq) \rightarrow S_4O_6^{2-} (aq) + 2 I^- (aq)$.

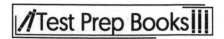

- **Acid-Base Reactions**: Involves a reaction between an acid and a base, which produces a salt and water. For example, $HBr + NaOH \rightarrow NaBr + H_2O$.

- **Combustion Reactions**: A hydrocarbon (a compound composed of only hydrogen and carbon) reacts with oxygen (O_2) to form carbon dioxide (CO_2) and water (H_2O). For example, $CH_4 + 2O_2 \rightarrow CO_2 + 2H_2O$.

Balancing Chemical Reactions

Chemical reactions are conveyed using chemical equations. **Chemical equations** must be balanced with equivalent numbers of atoms for each type of element on each side of the equation. Antoine Lavoisier, a French chemist, was the first to propose the **Law of Conservation of Mass** for the purpose of balancing a chemical equation. The law states, "Matter is neither created nor destroyed during a chemical reaction."

The **reactants** are located on the left side of the arrow, while the **products** are located on the right side of the arrow. **Coefficients** are the numbers in front of the chemical formulas. **Subscripts** are the numbers to the lower right of chemical symbols in a formula. To tally atoms, one should multiply the formula's coefficient by the subscript of each chemical symbol. For example, the chemical equation $2 H_2 + O_2 \rightarrow 2H_2O$ is balanced. For H, the coefficient of 2 multiplied by the subscript 2 = 4 hydrogen atoms. For O, the coefficient of 1 multiplied by the subscript 2 = 2 oxygen atoms. Coefficients and subscripts of 1 are understood and never written. When known, the form of the substance is noted with (g) for gas, (s) for solid, (l) for liquid, or (aq) for aqueous.

Stoichiometry

Stoichiometry investigates the quantities of chemicals that are consumed and produced in chemical reactions. Chemical equations are made up of reactants and products; stoichiometry helps elucidate how the changes from reactants to products occur, as well as how to ensure the equation is balanced.

Limiting Reactants

Chemical reactions are limited by the amount of starting material—or reactants—available to drive the process forward. The reactant that is present in the smallest quantity in a reaction is called the **limiting reactant**. The limiting reactant is completely consumed by the end of the reaction. The other reactants are called **excess reactants**. For example, gasoline is used in a combustion reaction to make a car move and is the limiting reactant of the reaction. If the gasoline runs out, the combustion reaction can no longer take place, and the car stops.

Reaction Yield

The quantity of product that should be produced after using up all of the limiting reactant can be calculated and is called the **theoretical yield** of the reaction. Since the reactants do not always act as they should, the actual amount of resulting product is called the **actual yield.** The actual yield is divided by the theoretical yield and then multiplied by 100 to find the percent yield for the reaction.

Solution Stoichiometry

Solution stoichiometry deals with quantities of solutes in chemical reactions that occur in solutions. The quantity of a solute in a solution can be calculated by multiplying the molarity of the solution by the volume. Similar to chemical equations involving simple elements, the number of moles of the elements that make up the solute should be equivalent on both sides of the equation.

When the concentration of a particular solute in a solution is unknown, a titration is used to determine that concentration. In a titration, the solution with the unknown solute is combined with a standard

solution, which is a solution with a known solute concentration. The point at which the unknown solute has completely reacted with the known solute is called the equivalence point. Using the known information about the standard solution, including the concentration and volume, and the volume of the unknown solution, the concentration of the unknown solute is determined in a balanced equation. For example, in the case of combining acids and bases, the equivalence point is reached when the resulting solution is neutral. HCl, an acid, combines with NaOH, a base, to form water, which is neutral, and a solution of Cl⁻ ions and Na⁺ ions. Before the equivalence point, there are an unequal number of cations and anions and the solution is not neutral.

Reaction Rates and Equilibrium

Reaction Rates

The **rate of a reaction** is the measure of the change in concentration of the reactants or products over a certain period of time. Many factors affect how quickly or slowly a reaction occurs, such as concentration, pressure, or temperature. As the concentration of a reactant increases, the rate of the reaction also increases, because the frequency of collisions between elements increases. High-pressure situations for reactants that are gases cause the gas to compress and increase the frequency of gas molecule collisions, similar to solutions with higher concentrations. Reactions rates are then increased with the higher frequency of gas molecule collisions. Higher temperatures usually increase the rate of the reaction, adding more energy to the system with heat, and increasing the frequency of molecular collisions.

Equilibrium

Equilibrium is described as the state of a system when no net changes occur. Chemical equilibrium occurs when opposing reactions occur at equal rates. In other words, the rate of reactants forming products is equal to the rate of the products breaking down into the reactants — the concentration of reactants and products in the system does not change. Although the concentrations are not changing, the forward and reverse reactions are likely still occurring. This type of equilibrium is called a **dynamic equilibrium**. In situations where all reactions have ceased, a static equilibrium is reached. Chemical equilibriums are also described as homogeneous or heterogeneous. **Homogeneous equilibrium** involves substances that are all in the same phase, while **heterogeneous equilibrium** means the substances are in different phases when equilibrium is reached.

When a reaction reaches equilibrium, the conditions of the equilibrium are described by the following equation, based on the chemical equation aA + bB ↔ cC + dD:

$$K_c = \frac{[C]^c [D]^d}{[A]^a [B]^b}$$

This equation describes the law of mass action. It explains how the reactants and products react during dynamic equilibrium. K_c is the equilibrium constant, and it is obtained when molarity values are put into the equation for the reactants and products. It is important to note that K_c is only dependent on the stoichiometry of the equation. If K_c is greater than 1, the equilibrium occurs when there are more products generated; the equilibrium lies to the right. If K_c is less than 1, the equilibrium occurs when there are more reactants generated and the equilibrium is to the left.

Similar to finding K_c, the quantity of reactants and products, as well as the direction of the reaction, can be determined at any point of time by finding Q, the reaction quotient. Q_c is substituted for the K_c in the equation above. If Q is less than K, the concentration of the reactants is too large, and the concentration

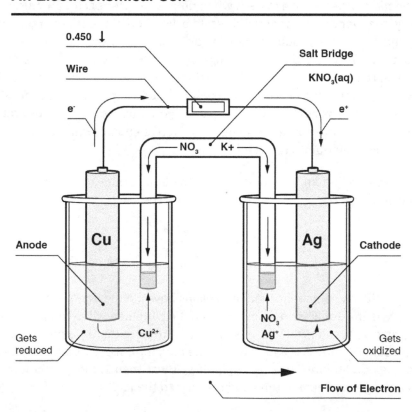

of the products is too small, so the reaction must move from left to right to achieve equilibrium. If Q is equal to K, the system is at equilibrium. If Q is greater than K, the concentration of the products is too large, and the concentration of the reactants is too small; the reaction must move from right to left to reach equilibrium.

Oxidation-Reduction Reactions

Oxidation-reduction reactions, or **redox reactions**, are chemical reactions in which electrons are transferred from one compound to another. A helpful mnemonic device to remember the basic properties of redox reactions is "LEO the lion says GER", in which "LEO" means Lose Electrons Oxidation and "GER" stands for Gain Electrons Reduction. A **reducing agent** is an electron donor, and an **oxidizing agent** is an electron acceptor.

An example of an oxidation-reduction reaction is the **electrochemical cell**, which is the basis for batteries. The electrochemical cell comprises two different cells, each equipped with a separate conductor, and a salt bridge. The **salt bridge** isolates reactants but maintains the electric current. The **cathode** is the electrode that is reduced, and the **anode** is the electrode that is oxidized. **Anions** and **cations** carry electrical current within the cell, and electrons carry current within the electrodes. The voltmeter in the picture below shows the difference in potential between the two cells. As this redox reaction moves toward equilibrium, the potential will decrease and approach 0.000V. In this example, silver is reduced while copper is oxidized.

An Electrochemical Cell

Solubility

Solubility is defined as the ability of a particular solute to dissolve in a solvent. It is usually measured in grams or moles of solute per a known volume of solution. Solubility of gases in a solution is influenced by pressure. Henry's Law states the amount of dissolved gas in a solution is directly proportional to the pressure of the solution. Solubility is also influenced by temperature; increasing temperature will speed up the rate at which most solids dissolve. When a solution reaches the point where no more solute is dissolving—often indicated by the precipitation of solid—it is called a **saturated solution**. In the case of gas solubility, a temperature increase will decrease gas solubility in water. Solubility of solids may also be impacted by common ion effects. Two compounds housing the same dissociating cation, combined in solution, will often significantly reduce the solubility of the material being evaluated. Physical agitation—such as stirring, sonication, or mixing—may speed up the dissolving process by increasing the molecular interactions of solute and solvent.

Precipitation Titrations

A titration in which the end point is determined by formation of a precipitate is called a **precipitation titration**. The most common precipitation titrations are argentometric (*Argentum* means silver in Latin), which involves a silver ion, with silver nitrate ($AgNO_3$) being the titrant. Silver nitrate is a very useful titrant because it reacts very quickly. All products in these titrimetric precipitations are silver salts. A very simple argentometric precipitation titration is often performed to determine the amount of chlorine in seawater, as the chlorine anion is present most abundantly. The reaction is as follows:

$$AgNO_3 \text{ (aq)} + NaCl \text{ (aq)} \rightarrow AgCl \text{ (s)} + NaNO_3 \text{ (aq)}$$

Silver chloride is almost completely insoluble in water, so it precipitates out of solution. A selective indicator, potassium chromate (K_2CrO_4), is added to determine when the reaction is complete. When all of the chloride ions from the sample have precipitated, silver ions then react with chromate ions to form Ag_2CrO_4, which is orange in color. The volume of the sample, amount of titrant required to complete the precipitation, molar mass of the reacting species, and the density of the sample are sufficient information to calculate the concentration of chloride ions.

Titration curves are useful in providing a graphical account of what occurs before, during, and after the equivalence point. Argentometric titration curves are sigmoidal when the concentration function is plotted on the y-axis and the reagent volume is plotted on the x-axis.

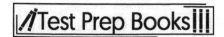

An example is shown below:

Sample Titration Curve

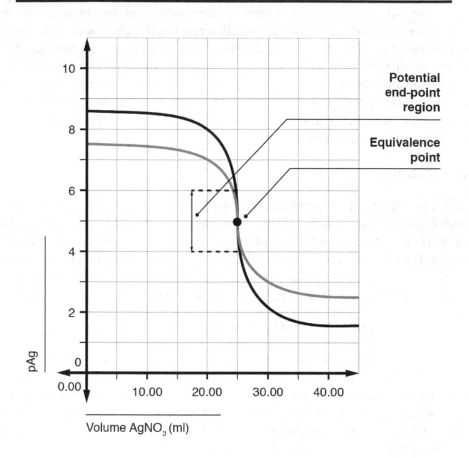

Volume AgNO$_3$ (ml)

Note that at the equivalence point, there is a sharp drop in reagent concentration, and then beyond the equivalence point, the concentration decreases gradually. Around the equivalence point, small volumes of titrant added cause dramatic changes in the concentration of analyte and reagent. At equivalence, the molar concentrations of sample cation/titrant anion and sample anion/titrant cation are equal.

Valence Bonds, Hybridization, and Molecular Orbitals

Valence Bond Theory

In **valence bond theory**, electrons are treated in a quantum mechanics fashion. Valence bond theory is good for describing the shapes of covalent compounds. Instead of electrons being pinpointed to an exact location, it's possible to map out a region of space—known as **atomic orbitals**—that the electron may inhabit. In valence bond theory, electrons are treated as excitations of the **electron field**, which exist everywhere. When energy is given to an electron field, electrons are said to exist inside a **wave function**—a mathematical function describing the probability that an electron is in a certain place at any given time. Standing waves are created when energy is given to a wave function. Electrons function as standing waves around a nucleus.

There are two things to consider in valence bond theory: the overlapping of the orbitals and the potential energy changes in a molecule as the atoms get closer or further apart. For example, the Lewis structure of H_2 is H – H, but this doesn't describe the strength of the bond. The electron cloud in the nucleus of one atom is interacting with the cloud of electrons in the nucleus of another atom. If the potential energy is mapped out as a function of distance between the two atoms, something like this is obtained:

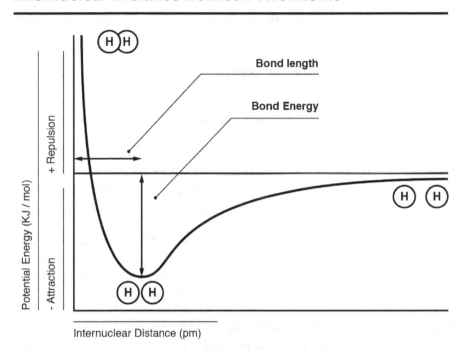

Potential Energy versus Internuclear Distance Between Two Atoms

The atoms approach this minimum value at a distance known as the **bond energy.** The link that corresponds to this is the equilibrium bond length—a happy medium corresponding to a minimum in potential energy. This minimum of energy is achieved when the atomic orbitals are overlapping and so contain two **spin-paired electrons**—when the two atoms' orbitals are overlapping each other. A greater amount of energy is needed to separate those two atoms when they are overlapped.

Molecular Orbitals

Atomic orbitals explain the behavior of a single electron or pairs of electrons in an atom. They are regions of space in which the electrons are more likely to spend their time. Every orbital can contain two electrons and is at its lowest energy when it has two electrons. One electron spins up, and one spins down. As mentioned, the standard atomic orbitals are known as *s*, *p*, *d*, and *f* orbitals.

- The simplest of the orbitals is the **s orbital**—the inner orbital of any atom or the outer orbital for light molecules, such as hydrogen and helium. The s orbital is spherical in shape and can contain two electrons.

- After the first and second s orbitals are filled, the **p orbitals** are filled. There are three p orbitals, one on each of the x, y, and z axes. Each p orbital can contain two electrons for a total of six electrons.

- After the p orbitals are filled, the **d orbitals** are filled next. There are five d orbitals, for a total of ten electrons.

- The **f orbitals** are next; there are seven f orbitals, which can contain a total of fourteen electrons.

Each shell/energy level has an increasing number of subshells available to it:

- The first shell only has the 1s subshell, so it has two electrons.
- The second shell has the 2s and 2p subshells, so it has (2 + 6) eight electrons.
- The third shell has the 3s, 3p and 3d subshells, so it has (2 + 6 + 10) 18 electrons.
- The fourth shell has the 4s, 4p, 4d and 4f subshells, so it has (2 + 6 + 10 + 14) 32 electrons.

To find the maximum number of electrons per shell, the formula $2n^2$ is used, where n is the shell number. For example, elements in the third period have three subshells—space for up to 18 electrons—but will only have up to eight valence electrons. This is because the *3d* orbitals are not filled—i.e., the elements from the third period do not completely fill their third shell.

Any element in the periodic table can be written in terms of its electron configuration. For instance, Calcium (Ca), which is in the fourth period on the periodic table and has an atomic number of 20, can be written as $1s^2 2s^2 2p^6 3s^2 3p^6 4s^2$. However, it's important to remember that the transition metals do not follow this rule. This is because quantum energy level rules allow for some of their shells to remain unfilled. For example, the transition metal scandium (Sc), which has an atomic number of 21, has the electron configuration $1s^2 2s^2 2p^6 3s^2 3p^6 3d^1 4s^2$, and the *d* subshell is not filled.

In **molecular orbital theory**, the assumption is that bonding, non-bonding, and anti-bonding orbitals, which have different energies, are formed when atoms are brought together. For N atomic orbitals in a molecule, the assumed result would be N molecular orbitals, which can be described by wave functions.

For example, for a molecule that has two atomic orbitals, two molecular orbitals must be formed: one bonding and one anti-bonding. The molecular orbitals would be separated by a certain energy. A molecule that has three atomic orbitals would form one bonding, one non-bonding, and one anti-bonding molecular orbital. A molecule that has ten atoms would form five anti-bonding and five bonding molecular orbitals.

Sigma bonds (σ bonds) are formed by direct overlapping of atomic orbitals, and they are the strongest type of covalent bond. Sigma bonds are symmetrical around the bond axis. Common sigma bonds—where z is the axis of the bond—are s+s, $p_z + p_z$, s+p_z, and $d_z{}^2 + d_z{}^2$.

In contrast, **pi bonds (π bonds)** are usually weaker than sigma bonds and are a type of covalent bond where two ends of one p-orbital overlap the two ends of another p-orbital. D-orbitals can also form pi bonds.

Generally, **single bonds** are sigma bonds, and multiple bonds consist of one sigma bond plus one pi bond. A **double bond** is one sigma bond plus one pi bond. A **triple bond** consists of one sigma bond and two pi bonds. For example, ethylene has **delta bonds (δ bonds)** that are formed from four ends of one atomic orbital overlapping with four ends of another atomic orbital.

Hybridization

Sometimes, the *s*, *p*, *d*, and *f* orbitals do not fully explain where an electron will be at any given time. This is where hybrid orbitals come in. **Hybrid orbitals** are combinations of the standard atomic orbitals. For instance, if the *s* and *p* orbitals hybridize, instead of being two different kinds of orbitals, they become four identical orbitals. When the *s* orbital hybridizes with all three *p* orbitals, it's called *sp*^{3} hybridization, and it forms a tetrahedral shape. This is the type of hybridization that occurs in H_2O. The description used depends on the properties of the compound, including its numbers of lone pairs of electrons.

Bond Energies

The **bond energy** can be described as the amount of energy required to break apart one mole of covalently-bonded gases. Bond energies are measured in kilojoules per mole of bonds (kJ/mol). The **mole** is a unit of measurement that is commonly used in chemistry. One mole of something is equal to Avogadro's number: $6.02214076 \times 10^{23}$. To calculate the number of moles of something, some simple formulas can be used:

Moles = mass (g) / Relative mass (grams per mole).

Example: How many moles are there in 30 grams of helium?

On the periodic table, Helium's relative mass is approximately two. Using this information and the formula yields:

Moles = 30 / 2 = 15 moles

During a chemical reaction, some bonds are broken, and some are formed. Bonds do not break or form spontaneously—they require energy to be added or released. The energy needed to break a bond is the **bond energy**. As mentioned, generally, the shorter the bond length, the greater the bond energy.

When atoms combine to make bonds and form a compound, energy is always released, normally as heat. Certain types of bonds have similar bond energies, despite each molecule being different. For example, all C-H bonds will have a value of roughly 413 kJ/mol. There are reference lists available with average bond energies that have been calculated for various atoms bonded together.

Using bond energies, it is possible to calculate the **enthalpy change** within a system. When a chemical reaction occurs, there will always be an accompanying change in energy. Energy is released to make bonds; conversely, energy is required to break bonds.

- Some reactions are **exothermic**—where energy is released during the reaction, usually in the form of heat—because the energy of the products is lower than the energy of the reactants. In exothermic reactions, energy can be thought of as a **product**.

- Some reactions are **endothermic**—where energy is absorbed from the surroundings because the energy of the reactants is lower than the energy of the products. In endothermic reactions, energy can be thought of as a **reactant**.

It is possible to look at the two sides of a chemical reaction and work out whether energy is gained or lost during the formation of the products, thus determining whether the reaction is exothermic or endothermic. Here is an example:

Two moles of water forming two moles of hydrogen and one mole of oxygen:

$$2H_2O(g) \rightarrow 2H_2 + O_2(g)$$

The sum on the reactant's side (2 moles of water) is equal to four sets of H-O bonds, which is 4 x 460 kJ/mol = 1840 kJ/mol. This is the input.

The sum on the product's side is equal to 2 moles of H-H bonds and 1 mole of O=O bonds, which is 2 x 436.4 kJ/mol and 1 x 498.7 kJ/mol = 1371.5 kJ/mol. This is the output.

The total energy difference is 1840 − 1371.5 = +468.5 kJ/mol.

The energy difference is positive, which means that the reaction is endothermic, and the reaction will require energy (i.e., heat) to be carried out.

Covalent and van der Waals Radii

The **covalent radius (rcov)** is the length of one half of the bond length when two atoms of the same kind are bonded through a single bond in a neutral molecule. The sum of two covalent radii from atoms that are covalently bonded should, theoretically, be equal to the covalent bond length:

$$rcov(AB) = r(A) + r(B)$$

The van der Waals radius, rv, can be defined as half the distance between the nuclei of two non-bonded atoms of the same element when they are as close as possible to each other without being in the same molecule or being covalently bonded.

The covalent radii changes depending on the environment that an atom is in and whether it is single, double, or triple bonded, but average values exist for use in calculations, for example.

The van der Waals radii also changes based on the intermolecular forces present, but average values also exist. These values are useful for determining how closely molecules can pack into a solid.

Physics

Energy

Energy may be defined as the capacity to do work. It can be transferred between objects, comes in a multitude of forms, and can be converted from one form to another. The **Law of Conservation of Energy** states, "Energy can neither be created nor be destroyed."

Kinetic Energy and Potential Energy

The two primary forms of energy are kinetic energy and potential energy. **Kinetic energy** involves the energy of motion. Kinetic energy can be calculated using the equation: $KE = 0.5mv^2$, where m is the mass and v is the velocity of the object.

Potential energy represents the energy possessed by an object by virtue of its position. Potential energy may be calculated using the equation: $PE = mgh$, where m is the mass, g is the acceleration of gravity, and h is the height of the object.

Other Forms of Energy

The following are forms of kinetic energy:

- **Radiant energy**: Represents electromagnetic energy, which usually travels in waves or particles. Examples of radiant energy include visible light, gamma rays, X-rays, and solar energy.

- **Thermal energy**: Refers to the vibration and movement of molecules and atoms within a substance. It is also known as heat. When an object is heated, its molecules and atoms move and collide with each other more quickly. Examples of thermal energy include geothermal energy from the Earth, heated swimming pools, baking in an oven, and the warmth of a campfire on a person's skin.

- **Sound energy**: Represents the movement of energy through a substance, such as water or air, in waves. Vibrations cause sound energy. Examples of sound energy include a person's voice, clapping, singing, and musical instruments.

- **Electrical energy**: Utilizes charged particles called electrons moving through a wire. Some examples of electrical energy include lightning, batteries, alternating current (AC), direct current (DC), and static electricity.

- **Mechanical energy**: Represents the energy derived from the movement of objects. It is also called motion energy. Examples of mechanical energy include wind, flight of an airplane, electrons orbiting an atom's nucleus, and running.

The following are forms of potential energy:

- **Chemical energy**: Represents the energy stored in the bonds of molecules and atoms. Chemical energy usually undergoes conversion to thermal energy. Examples of chemical energy include petroleum, coal, and natural gas.

- **Elastic energy**: Represents the energy stored in an object as its volume or shape is distorted. It is also known as stored mechanical energy. Examples of elastic energy include stretched rubber bands and coiled springs.

- **Gravitational energy**: Refers to the energy stored as a result of an object's place or position. Increases in height and mass translate into increased gravitational energy. Examples of systems using gravitational energy include rollercoasters and hydroelectric power.

- **Nuclear energy**: Refers to the energy stored in an atom's nucleus. Vast amounts of energy can be released through the combination or splitting of nuclei. An example of nuclear energy is the nucleus of the element uranium.

Mechanics

Description of Motion in One and Two Dimensions

The description of motion is known as **kinetics**, and the causes of motion are known as **dynamics**. Motion in one dimension is a **scalar** quantity. It consists of one measurement such as length (length/distance is also known as **displacement**), speed, or time. Motion in two dimensions is a **vector** quantity. This is a speed with a direction, or **velocity**.

Velocity is the measure of the change in distance over the change in time. All vector quantities have a direction that can be relayed through the sign of an answer, such as -5.0 m/s or +5.0 m/s. The objects registering these velocities would be traveling in opposite directions, where the change in distance is denoted by Δx and the change in time is denoted by Δt:

$$v = \frac{\Delta x}{\Delta t}$$

Acceleration is the measure of the change in an object's velocity over a change in time, where the change in velocity, $v_2 - v_1$, is denoted by Δv and the change in time, $t_1 - t_2$, is denoted by Δt:

$$a = \frac{\Delta v}{\Delta t}$$

The **linear momentum,** p, of an object is the result of the object's mass, m, multiplied by its velocity, v, and is described by the equation:

$$p = mv$$

This aspect becomes important when one object hits another object. For example, the linear momentum of a small sports car will be much smaller than the linear momentum of a large semi-truck. Thus, the semi-truck will cause more damage to the car than the car to the truck.

Newton's Three Laws of Motion

Newton's Three Laws of Motion

Sir Isaac Newton summarized his observations and calculations relating to motion into three concise laws:

First Law of Motion: Inertia

This law states that an object in motion tends to stay in motion or an object at rest tends to stay at rest unless the object is acted upon by an outside force.

For example, a rock sitting on the ground will remain in the same place, unless it is pushed or lifted from its place.

This law also includes the relation of weight to gravity and force between objects relative to the distance separating them.

$$Weight = G\frac{Mm}{r^2}$$

Where G is the gravitational constant, M and m are the masses of the two objects, and r is the distance separating the two objects.

Second Law of Motion: F = ma

This law states that the force on a given body is the result of the object's mass multiplied by the acceleration acting upon the object. For objects falling on Earth, acceleration is caused by gravitational force ($9.8\ m/s^2$).

Third Law of Motion: Action-Reaction

This law states that for every action there is an equal and opposite reaction. For example, if a person punches a wall, the wall exerts a force back on the person's hand equal and opposite to his or her punching force. Since the wall has more mass, it absorbs the impact of the punch better than the person's hand.

Mass, Weight, and Gravity

Mass is a measure of how much of a substance exists, or how much inertia an object has. The mass of an object does not change based on the object's location, but the weight of an object does vary with its location.

For example, a 15-kg mass has a weight that is determined by acceleration from the force of gravity here on Earth. However, if that same 15 kg mass was weighed on the moon, it would weigh much less, since the acceleration force from the moon's gravitational force is approximately one-sixth of that on Earth.

Weight = mass × acceleration of gravity

$$W_{Earth} = 15\ kg \times 9.8\ m/s^2 > W_{Moon} = 15\ kg \times 1.62\ m/s^2$$

$$W_{Earth} = 147N > W_{Moon} = 24.3N$$

Analysis of Motion and Forces

Projectile motion describes the path of an object in the air. Generally, it is described by two-dimensional movement, such as a stone thrown through the air. This activity takes the shape of a parabolic curve. However, the definition of projectile motion also applies to free fall, or the non-arced motion of an object in a path straight up and/or straight down. When an object is thrown horizontally, it is subject to the same influence of gravity as an object that is dropped straight down. The farther the projectile motion, the farther the distance of the object's flight.

Friction is a force that opposes motion. It can be caused by several materials; there are even frictions caused by air or water. Whenever two differing materials touch, rub, or pass by each other, this can create friction, or an oppositional force. To move an object across a floor, the force exerted on the object must overcome the frictional force keeping the object in place. Friction is also why people can walk on surfaces. Without the oppositional force of friction to a shoe pressing on the floor, a person would not be able to grip the floor to walk—similar to the challenges of walking on ice. Without friction, shoes slip and are unable to help people push forward and walk.

When calculating the effects of objects colliding with each other, several things are important to remember. One of these is the definition of **momentum**: the mass of an object multiplied by the object's velocity. It is expressed by the following equation:

$$p = mv$$

Here, p is equal to an object's momentum, m is equal to the object's mass, and v is equal to the object's velocity.

Another important thing to remember is the principal of the **conservation of linear momentum**. The total momentum for objects in a situation will be the same before and after a collision. There are two primary types of collisions: **elastic** and **inelastic**. In an elastic collision, the objects collide and then travel in different directions. During an inelastic collision, the objects collide and then stick together in their final direction of travel. The total momentum in an elastic collision is calculated by using the following formula:

$$m_1 v_1 + m_2 v_2 = m_1 v_1 + m_2 v_2$$

Here, m_1 and m_2 are the masses of two separate objects, and v_1 and v_2 are their velocities, respectively.

The total momentum in an inelastic collision is calculated by using the following formula:

$$m_1 v_1 + m_2 v_2 = (m_1 + m_2) v_f$$

Here, v_f is the final velocity of the two masses after they stick together post-collision.

Example:

> If two bumper cars are speeding toward each other and collide head-on, they are designed to bounce off of each other and head in different directions. This would be an elastic collision.

> If real cars were speeding toward each other and collided head-on, there is a good chance their bumpers might get caught together and their subsequent direction of travel would be in the same direction together.

Circular motion takes place around an **axis,** which is an invisible line around which an object can rotate. This type of motion can be observed in the movements of a toy top. There is actually a point (or rod) through the center of a toy top on which the top can be observed as the spinning point for the top. This rod is called the axis.

When objects move in a circle by spinning on their own axis, or from being tethered around a central point (also an axis), they exhibit **circular motion**. In many ways, circular motion is similar to linear (straight line) motion. One difference is, when spinning an object on or around an axis, a force is created that feels like it is pushing out from the center of the circle. In reality, the force is actually pulling into the center of the circle, and the reactionary force is what is creating the feeling of pushing out. The

inward force is the real force and is called **centripetal** force. The outward, or reactionary, force is called **centrifugal** force. The reactionary force is not the real force—it just feels like it is there. This has also been referred to as a **fictional** force. The true force is the one pulling inward, or the centripetal force. The terms *centripetal* and *centrifugal* are often mistakenly interchanged.

Example:

> A traditional upright-style washing machine spins a load of laundry to remove the water from the load. The machine spins a barrel with holes in a circle at a high rate of speed. A force is pulling in toward the center of the circle (centripetal force). At the same time, the force reactionary to the centripetal force is pressing the laundry up against the outer sides of the barrel, thus pushing the water through the small holes that line the outer wall of the barrel.

An object moving in a circular motion also has momentum. In a circular motion, this is called **angular momentum**. It is determined by rotational inertia, rotational velocity, and the distance of the mass from the axis of rotation, or center of rotation.

Objects exhibiting circular motion also demonstrate the conservation of angular momentum. This means that the angular momentum of a system is always constant, regardless of the placement of the mass. Rotational inertia can be affected by how far the mass of the object is placed with respect to the center of rotation (axis of rotation). The farther the mass is from the center of rotation, the slower the rotational velocity. Conversely, if the mass is closer to the center of rotation, the rotational velocity increases. A change in one affects the other, thus conserving the angular momentum. This holds true as long as no external forces act on the system.

Example:

> When ice skaters are spinning on one ice skate, they can extend their arms to slow their rotational velocity. However, when skaters bring their arms in close to their bodies (or shorten the distance between the mass and the center of rotation), their rotational velocity increases, and they will spin much faster. Some skaters extend their arms straight above their heads, which causes an extension of the axis of rotation, thus removing any distance between the mass and the center of rotation and maximizing their rotational velocity. In other words, they spin extremely fast.

The center of mass is the point that provides the average location for the total mass of a system. The word "system" can apply to just one object/particle or to many. The center of mass for a system can be calculated by finding the average of the mass of each object and multiplied by its distance from an origin point using the following formula:

$$x_{center\ of\ mass} = \frac{m_1 x_1 + m_2 x_2}{m_1 + m_2}$$

In this case, *x* is the distance from the point of origin for the center of mass and each respective object, and *m* is the mass of each object.

To calculate for more than one object, the pattern can be continued by adding additional masses and their respective distances from the origin point.

Simple Machines

A **simple machine** is a mechanical device that changes the direction or magnitude of a force. There are six basic types of simple machines: lever, wedge, screw, inclined plane, wheel and axle, and pulley.

Here is how each type works and an example:

- A **lever** helps lift heavy items higher with less force, such as a crowbar lifting a large cast iron lid.

- A **wedge** helps apply force to a specific area by focusing the pressure, such as an axe splitting a tree.

- An **inclined plane** helps move heavy items up vertical distances with less force, such as a loading dock ramp.

- A **screw** is an inclined plane wrapped around an axis and allows more force to be applied by extending the distance of the plane, such as a screw being turned into a piece of wood rather than hitting a nail into the wood.

- A **wheel and axle** allows the use of rotational force around an axis to assist with applying force, such as a wheelbarrow making it easier to haul large loads by employing a wheel and axle at the front.

- A **pulley** is an application of a wheel and axle with the addition of cords or ropes and helps move objects vertically, such as pulling a bucket out of a well is easier with a pulley and ropes.

Here's some visual representations:

wheel and axle

pulley

wedge

pry bar

screw

inclined plane

Using a simple machine employs an advantage to the user. This is referred to as the **mechanical advantage**. It can be calculated by comparing the force input by the user to the simple machine with the force output from the use of the machine (also displayed as a ratio).

$$Mechanical\ Advantage\ = \frac{output\ force}{input\ force}$$

$$MA\ = \frac{F_{out}}{F_{in}}$$

When using a lever, it can be helpful to calculate the torque, or circular force, necessary to move something. This is also employed when using a wrench to loosen a bolt.

$$Torque\ =\ F\ \times\ distance\ of\ lever\ arm\ from\ the\ axis\ of\ rotation$$

$$T\ =\ F\ \times\ d$$

Electricity and Magnetism

Electrical Nature of Common Materials

Generally, an atom carries no net charge because the positive charges of the protons in the nucleus balance the negative charges of the electrons in the outer shells of the atom. This is considered to be electrically neutral. However, electrons are the only portion of the atom known to have the freedom to "move," and this can cause an object to become electrically charged. This happens either through a gain or a loss of electrons. Electrons have a negative charge, so a gain creates a negative charge for the object. On the contrary, a loss of electrons creates a positive charge for the object. This charge can also be focused on specific areas of an object, causing a notable interaction between charged objects. For example, if a person rubs a balloon on a carpet, the balloon transfers some of its electrons to the carpet. So, if a person were to hold the balloon near his or her hair, the electrons in the "neutral" hair would make the hair stand on end. This is due to the electrons wanting to fill the deficit of electrons on the balloon. Unless electrically forced into a charged state, most natural objects in nature tend toward reestablishing and maintaining a neutral charge.

When dealing with charges, it is easiest to remember that **like charges _repel_** each other and **opposite charges attract** each other. Therefore, negatives and positives attract, while two positives or two negatives repel each other. Similarly, when two charges come near each other, they exert a force on one another. This is described through **Coulomb's Law**:

$$F\ =\ k\frac{q_1 q_2}{r^2}$$

Where F is equal to the force exerted by the interaction, k is a constant ($k = 8.99$ x 10^9 N m^2/C^2), q_1 and q_2 are the measure of the two charges, and r is the distance between the two charges.

When materials readily transfer electricity or electrons, or can easily accept or lose electrons, they are considered to be good conductors. The transferring of electricity is called **conductivity**. If a material does not readily accept the transfer of electrons or readily loses electrons, it is considered to be an insulator. For example, copper wire easily transfers electricity because copper is a good conductor. However, plastic does not transfer electricity because it is not a good conductor. In fact, plastic is an insulator.

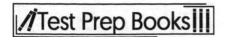

Basic Electrical Concepts

In an electrical circuit, the flow from a power source, or the voltage, is "drawn" across the components in the circuit from the positive end to the negative end. This flow of charge creates an electric current (I), which is the time (t) rate of flow of net charge (q). It is measured with the formula:

$$I = \frac{q}{t}$$

Current is measured in **amperes** (amps). There are two main types of currents: (1) **direct current** (DC) — a one directional flow of charges through a circuit; and (2) **alternating current** (AC) — a circuit with a changing directional flow of charges or magnitude.

Every circuit will show a loss in voltage across its conducting material. This loss of voltage is from resistance within the circuit and can be caused by multiple factors, including resistance from wiring and components such as light bulbs and switches. To measure the resistance in a given circuit, **Ohm's law** is used:

$$Resistance = \frac{Voltage}{current}$$

$$R = \frac{V}{I}$$

Resistance (R) is measured in **Ohms** (Ω).

Components in a circuit can be wired **in series** or **in parallel**. If the components are wired in series, a single wire connects each component to the next in line. If the components are wired in parallel, two wires connect each component to the next. The main difference is that the voltage across those in series is directly related from one component to the next. Therefore, if the first component in the series becomes inoperable, no voltage can get to the other components. Conversely, the components in parallel share the voltage across each other and are not dependent on the component wired prior to the next to allow the voltage across the wire.

To calculate the resistance of circuit components wired in series or parallel, the following equations are used:

Resistance in series:

$$R_{total} = R_1 + R_2 + R_3 + \cdots$$

Resistance in parallel:

$$R_{total} = \frac{1}{R_1} + \frac{1}{R_2} + \frac{1}{R_3} + \cdots$$

To make electrons move so they can carry their charge, a change in voltage must be present. On a small scale, this is demonstrated through the electrons traveling from the light switch to a person's finger. This might happen in a situation where a person runs his or her socks on a carpet, touches a light switch, and receives a small jolt from the electrons that run from the switch to the finger. This minor jolt is due to the deficit of electrons created by rubbing the socks on the carpet, and then the electrons going into the ground. The difference in charge between the switch and the finger caused the electrons to move.

If this situation were to be created on a larger and more sustained scale, the factors would need to be more systematic, predictable, and harnessed. This can be achieved through batteries/cells and generators. Batteries or cells have a chemical reaction that occurs inside, causing energy to be released and charges to be able to move freely. Batteries generally have nodes (one positive and one negative), where items can be hooked up to complete a circuit and allow the charge to travel freely through the item. Generators convert mechanical energy into electric energy using power and movement.

Basic Properties of Magnetic Fields and Forces

Consider two straight rods that are made from magnetic material. They will naturally have a negative end (pole) and a positive end (pole). These charged poles react just like any charged item: opposite charges attract and like charges repel. They will attract each other when arranged positive pole to negative pole. However, if one rod is turned around, the two rods will now repel each other due to the alignment of negative to negative and positive to positive. These types of forces can also be created and amplified by using an electric current. For example, sending an electric current through a stretch of wire creates an electromagnetic force around the wire from the charge of the current. This force exists as long as the flow of electricity is sustained. This magnetic force can also attract and repel other items with magnetic properties. Depending on the strength of the current in the wire, a greater or smaller magnetic force can be generated around the wire. As soon as the current is stopped, the magnetic force also stops.

Optics and Waves

Electromagnetic Spectrum

The movement of light is described like the movement of waves. Light travels with a wave front, has an amplitude (height from the neutral), a cycle or wavelength, a period, and energy. Light travels at approximately 3.00×10^8 m/s and is faster than anything created by humans.

Light is commonly referred to by its measured wavelengths, or the distance between two successive crests or troughs in a wave. Types of light with the longest wavelengths include radio, TV, and micro and infrared waves. The next set of wavelengths are detectable by the human eye and create the **visible spectrum**. The visible spectrum has wavelengths of 10^{-7} m, and the colors seen are red, orange, yellow, green, blue, indigo, and violet. Beyond the visible spectrum are shorter wavelengths, also called the **electromagnetic spectrum**, containing ultraviolet light, X-rays, and gamma rays. The wavelengths outside of the visible light range can be harmful to humans if they are directly exposed or exposed for long periods of time.

Basic Characteristics and Types of Waves

A **mechanical wave** is a type of wave that passes through a medium (solid, liquid, or gas). There are two basic types of mechanical waves: longitudinal and transverse.

A **longitudinal wave** has motion that is parallel to the direction of the wave's travel. It can best be shown by compressing one side of a tethered spring and then releasing that end. The movement travels in a bunching/un-bunching motion across the length of the spring and back.

A **transverse wave** has motion that is perpendicular to the direction of the wave's travel. The particles on a transverse wave do not move across the length of the wave, but instead oscillate up and down to create peaks and troughs.

A wave with a mix of both longitudinal and transverse motion can be seen through the motion of a wave on the ocean—with peaks and troughs, and particles oscillating up and down.

Mechanical waves can carry energy, sound, and light, but they need a medium through which transport can occur. An electromagnetic wave can transmit energy without a medium, or in a vacuum.

A more recent addition in the study of waves is the **gravitational wave**. Its existence has been proven and verified, yet the details surrounding its capabilities are still somewhat under inquiry. They are purported to be ripples that propagate as waves outward from their source and travel in the curvature of space/time. They are thought to carry energy in a form of radiant energy called gravitational radiation.

Basic Wave Phenomena

When a wave crosses a boundary or travels from one medium to another, certain actions occur. If the wave is capable of traveling through one medium into another medium, it experiences **refraction**. This is the bending of the wave from one medium to another due to a change in density and thus the speed of the wave. For example, when a pencil is sitting in half of a glass of water, a side view of the glass makes the pencil appear to be bent at the water level. What the viewer is seeing is the refraction of light waves traveling from the air into the water. Since the wave speed is slowed in water, the change makes the pencil appear bent.

When a wave hits a medium that it cannot penetrate, it is bounced back in an action called **reflection**. For example, when light waves hit a mirror, they are reflected, or bounced, off the mirror. This can cause it to seem like there is more light in the room, since there is "doubling back" of the initial wave. This same phenomenon also causes people to be able to see their reflection in a mirror.

When a wave travels through a slit or around an obstacle, it is known as **diffraction**. A light wave will bend around an obstacle or through a slit and cause what is called a **diffraction pattern**. When the waves bend around an obstacle, it causes the addition of waves and the spreading of light on the other side of the opening.

Dispersion is used to describe the splitting of a single wave by refracting its components into separate parts. For example, if a wave of white light is sent through a dispersion prism, the light appears as its separate rainbow-colored components, due to each colored wavelength being refracted in the prism.

When wavelengths hit boundaries, different things occur. Objects will absorb certain wavelengths of light and reflect others, depending on the boundaries. This becomes important when an object appears to be a certain color. The color of an object is not actually within that object, but rather, in the wavelengths being transmitted by that object. For example, if a table appears to be red, that means the table is absorbing all other wavelengths of visible light except those of the red wavelength. The table is reflecting, or transmitting, the wavelengths associated with red back to the human eye, and so it appears red.

Interference describes when an object affects the path of a wave, or another wave interacts with a wave. Waves interacting with each other can result in either **constructive interference** or **destructive interference**, based on their positions. With constructive interference, the waves are in sync with each other and combine to reinforce each other. In the case of deconstructive interference, the waves are out of sync and reduce the effect of each other to some degree. In **scattering**, the boundary can change the direction or energy of a wave, thus altering the entire wave. **Polarization** changes the oscillations of

a wave and can alter its appearance in light waves. For example, polarized sunglasses remove the "glare" from sunlight by altering the oscillation pattern observed by the wearer.

When a wave hits a boundary and is completely reflected, or cannot escape from one medium to another, it is called **total internal reflection**. This effect can be seen in the diamonds with a brilliant cut. The angle cut on the sides of the diamond causes the light hitting the diamond to be completely reflected back inside the gem, making it appear brighter and more colorful than a diamond with different angles cut into its surface.

The **Doppler effect** applies to situations with both light and sound waves. The premise of the Doppler effect is that, based upon the relative position or movement of a source and an observer, waves can seem shorter or longer than they are in actuality. When the Doppler effect is noted with sound, it warps the noise being heard by the observer. This makes the pitch or frequency seem shorter or higher as the source is approaching, and then longer or lower as the source is getting farther away. The frequency/pitch of the source never actually changes, but the sound in respect to the observer makes it seem like the sound has changed. This can be observed when a siren passes by an observer on the road. The siren sounds much higher in pitch as it approaches the observer and then lower after it passes and is getting farther away.

The Doppler effect also applies to situations involving light waves. An observer in space would see light approaching as being shorter wavelengths than the light is in actuality, causing it to seem blue. When the light wave gets farther away, the light would seem red because of the apparent elongation of the wavelength. This is called the **red-blue shift**.

Basic Optics

When reflecting light, a mirror can be used to observe a virtual (not real) image. A **plane mirror** is a piece of glass with a coating in the background to create a reflective surface. An image is what the human eye sees when light is reflected off the mirror in an unmagnified manner. If a **curved mirror** is used for reflection, the image seen will not be a true reflection. Instead, it will either be magnified or made smaller in its image than its actual size. Curved mirrors can also make the object appear closer or farther away than the actual distance the object is from the mirror.

Lenses can be used to refract or bend light to form images. Examples of lenses are the human eye, microscopes, and telescopes. The human eye interprets the refraction of light into images that humans understand to be actual size. **Microscopes** allow objects that are too small for the unaided human eye to be enlarged enough to be seen. **Telescopes** allow objects to be viewed that are too far away to be seen with the unaided eye. **Prisms** are pieces of glass that can have a wavelength of light enter one side and appear to be divided into its component wavelengths on the other side. This is due to the ability of the prism to slow certain wavelengths more than others.

Sound

Sound travels in waves and is the movement of vibrations through a medium. It can travel through air (gas), land, water, etc. For example, the noise a human hears in the air is the vibration of the waves as they reach the ear. The human brain translates the different frequencies (pitches) and intensities of the vibrations to determine what created the noise.

A tuning fork has a predetermined frequency because of the size (length and thickness) of its tines. When struck, it allows vibrations between the two tines to move the air at a specific rate. This creates a specific tone, or note, for that size of tuning fork. The number of vibrations over time is also steady for that tuning fork and can be matched with a frequency (the number of occurrences over time). All frequencies heard by the human ear are categorized by using frequency and are measured in Hertz (cycles per second).

The level of sound in the air is measured with sound level meters with a decibel (dB) scale. These meters respond to changes in air pressure caused by sound waves and measure sound intensity. One decibel is 1/10th of a *bel*, named after Alexander Graham Bell, inventor of the telephone. The decibel scale is logarithmic (not linear like a scale on a ruler), so it is measured in factors of 10. This means, for example, that a 10 dB increase with a sound meter is equal to a 10-fold increase in sound intensity.

Practice Questions

1. Which of the following is a type of boundary between two tectonic plates?
    a. Continental
    b. Oceanic
    c. Convergent
    d. Fault

2. Volcanic activity can occur in which of the following?
    a. Convergent boundaries
    b. Divergent boundaries
    c. The middle of a tectonic plate
    d. All of the above

3. The fact that the Earth is tilted as it revolves around the Sun creates which phenomenon?
    a. Life
    b. Plate tectonics
    c. Wind
    d. Seasonality

4. Where is most of the Earth's weather generated?
    a. The troposphere
    b. The ionosphere
    c. The thermosphere
    d. The stratosphere

5. What type of cloud is seen when looking at the sky during a heavy rainstorm?
    a. High-Clouds
    b. Altocumulus
    c. Stratus
    d. Nimbostratus

6. What is the largest planet in our solar system and what is it mostly made of?
    a. Saturn, rocks
    b. Jupiter, ammonia
    c. Jupiter, hydrogen
    d. Saturn, helium

7. Viruses belong to which of the following classifications?
    a. Domain Archaea
    b. Kingdom Monera
    c. Kingdom Protista
    d. None of the above

8. In which phase of mitosis does DNA replication occur?
    a. Anaphase
    b. Metaphase
    c. Telophase
    d. None of the above

9. Which of the following is true regarding the middle of a cell membrane?
    a. It is hydrophilic.
    b. It is hydrophobic.
    c. It is made of phosphate.
    d. It is made of cellulose.

10. What is the complement strand to this segment of DNA?

$$5' - AGTCCA - 3'$$

    a. $5' - TCAGGT - 3'$
    b. $5' - TGGACT - 3'$
    c. $3' - UGGACU - 5'$
    d. $3' - TGAGGT - 5'$

11. What are the lowest coefficients that will balance the following combustion equation?

$$\_\_C_2H_{10} + \_\_O_2 \rightarrow \_\_H_2O + \_\_CO_2$$

    a. 1:5:5:2
    b. 4:10:20:8
    c. 2:9:10:4
    d. 2:5:10:4

12. Which of the following represents a Punnett square for a child with a father who is Gg and mother who is Gg? (G = green eyes; g = gray eyes)

a.

	Father g	Father g
Mother G	Gg	Gg
Mother G	Gg	Gg

b.

	Mother g	Father g
Mother G	Gg	Gg
Father G	Gg	Gg

c.

	Father G	Father g
Mother G	GG	Gg
Mother g	Gg	gg

d.

	Father G	Father g
Mother g	Gg	gg
Mother g	gg	gg

13. All single-celled eukaryotic organisms belong to which classification?
    a. Archaea
    b. Monera
    c. Protista
    d. Kingdom

14. The somatic nervous system is responsible for which of the following?
    a. Breathing
    b. Thought
    c. Movement
    d. Fear

15. Which factor does NOT affect the solubility of a compound?
    a. Mass
    b. Chemical structure of the solute
    c. Temperature
    d. Common ion effects

16. What is used as the titrant in the commonly used argentometric titration?
    a. Silver chloride
    b. Calcium oxalate
    c. Silver sulfate
    d. Silver nitrate

17. Which statement is true regarding electrostatic charges?
    a. Like charges attract.
    b. Like charges repel.
    c. Like charges are neutral.
    d. Like charges neither attract or repel.

18. What is the molarity of a solution made by dissolving 4.0 grams of NaCl into enough water to make 120 mL of solution? The atomic mass of Na is 23.0 g/mol and Cl is 35.5 g/mol.
    a. 0.34 M
    b. 0.57 M
    c. 0.034 M
    d. 0.057 M

19. What is the current when a 3.0 V battery is wired across a lightbulb that has a resistance of 6.0 ohms?
    a. 0.5 A
    b. 18.0 A
    c. 0.5 J
    d. 18.0 J

20. What effect changes the oscillations of a wave and can alter its appearance in light waves?
    a. Reflection
    b. Refraction
    c. Dispersion
    d. Polarization

21. The Sun transfers heat to the Earth through space via which mechanism?
    a. Convection
    b. Conduction
    c. Induction
    d. Radiation

22. What is 45 ^{0}C converted to ^{0}F?
    a. 113 ^{0}F
    b. 135 ^{0}F
    c. 57 ^{0}F
    d. 88 ^{0}F

23. What type of chemical reaction produces a salt?
    a. Oxidation reaction
    b. Neutralization reaction
    c. Synthesis reaction
    d. Decomposition reaction

24. The Doppler effect applies to sound waves and light waves. What is its effect in space called?
    a. Red-black shift
    b. Red-blue shift
    c. Blue-black shift
    d. Blue-blue shift

25. Circular motion occurs around what?
    a. The center of mass
    b. The center of matter
    c. An elliptical
    d. An axis

Answer Explanations

1. C: Convergent plate boundaries occur where two tectonic plates collide together. The denser oceanic plate will drop below the continental plate in a process called subduction.

2. D: Volcanic activity can occur at both fault lines and within the area of a tectonic plate at areas called hot spots. Volcanic activity is more common at fault lines because of cracks that allow the mantle's magma to more easily escape to the surface.

3. D: This is the only answer choice created by Earth's tilt. The Earth rotates around the Sun at an axis of 23.5 degrees, which causes different latitudes to receive varying amounts of direct sunlight throughout the year.

4. A: Technically, the troposphere is a layer of the atmosphere where the majority of the activity that creates weather conditions experienced on Earth occurs. The ozone layer is in the stratosphere; this is also where airplanes fly.

5. D: Stratus clouds are also grey, but nimbostratus clouds are the low clouds that appear during stormy weather. The other choices are usually seen on fair-weather days.

6. C: Jupiter is the largest planet in the solar system, and it is primarily composed of hydrogen and helium. Ammonia is in much lower quantity and usually found as a cloud within Jupiter's atmosphere.

7. D: Viruses are not classified as living organisms. They are neither prokaryotic or eukaryotic; therefore, they don't belong to any of the answer choices.

8. D: DNA is replicated during the S-phase of interphase, which isn't considered part of mitosis. Mitosis is just the process of separation.

9. B: A cell membrane is a phospholipid bilayer, with the lipid or fat portion (the tails) composing the middle. Lipids are hydrophobic or water-fearing. Think about how difficult it is to mix oil (a lipid) and water. The phosphate portion is hydrophilic or water-loving and composes the surfaces of the membrane. Cellulose is only found in plant cells, which have a cell wall instead of a cell membrane.

10. B: This is tricky because you must pay attention to the direction of the strands. Although Choice *A* matches up with the question's strand, they cannot join together because the two strands must be going in opposite directions. Choice *C* is incorrect because it contains uracil, which is only found in RNA. Choice *D* is incorrect because it has a G where there should be a C.

11. C: 2:9:10:4. These are the coefficients that follow the law of conservation of matter. The coefficient times the subscript of each element should be the same on both sides of the equation.

12. C: Draw the square by placing the parents' alleles first and then match them up accordingly in the gray boxes.

	Mom G	Mom g
Dad G	GG	Gg
Dad g	Gg	gg

13. C: Kingdom Protista contains all the single-celled eukaryotic organisms. Kingdom Monera and Domain Archaea contain prokaryotes. Choice *D* is just a level of the classification system.

14. C: The somatic nervous system is the voluntary nervous system, responsible for voluntary movement. It includes nerves that transmit signals from the brain to the muscles of the body. Breathing is controlled by the autonomic nervous system. Thought and fear are complex processes that occur in the brain, which is part of the central nervous system.

15. A: Mass is the only listed factor that does not affect the solubility of a compound. Chemical structure (*B*) can influence solubility because "like dissolves like" i.e., a polar compound will not dissolve in a nonpolar solvent; temperature (*C*) can increase or decrease solubility; common ion effects (*D*) can decrease solubility if there are similarly charged dissociating agents in the solution; and pressure (*E*) affects the solubility of gases in solution.

16. D: Silver nitrate is the correct answer choice because the root of "argentometric" is **argentum,** which means "silver" in Latin. Silver nitrate is useful as a titrant because it reacts very quickly and, upon formation of a precipitate, can indicate the endpoint of the titration.

17. B: For charges, *like charges repel* each other and *opposite charges attract* each other. Negatives and positives will attract because the charges are opposite, while two positive charges or two negative charges will repel each other because the charges are the same. Charges influence each other, so the answers *C* and *D* are not reasonable.

18. B: To solve this, the number of moles of NaCl needs to be calculated:

First, to find the mass of NaCl, the mass of the molecule's atoms is totaled:

$$Na = 23.0\ g$$
$$Cl = +35.5\ g$$
$$58.5\ g\ NaCl$$

Next, the given mass of the substance is multiplied by one mole per total mass of the substance:

$$4.0\ g\ NaCl \times \frac{1\ mol\ NaCl}{58.5\ g\ NaCl} = 0.068\ mol\ NaCl$$

Finally, the moles are divided by the number of liters of the solution to find the molarity:

$$c = \frac{0.068 \text{ mol NaCl}}{0.120} = 0.57 \text{ M NaCl}$$

Choice *A* incorporates a miscalculation for the molar mass of NaCl. Choices *C* and *D* both fail to convert mL to liters (L), so they are incorrect by a factor of 10.

19. A: According to Ohm's Law: $V = IR$, so using the given variables:

$$3.0 \text{ V} = I \times 6.0 \text{ }\Omega$$

Solving for I:

$$I = 3.0 \text{ V}/6.0 \text{ }\Omega$$

$$I = 0.5 \text{ A}$$

Choice *B* shows a miscalculation in the equation by multiplying 3.0 V by 6.0 Ω, rather than dividing. Choices *C* and *D* are labeled with the wrong units; Joules measure energy, not current.

20. D: Polarization changes the oscillations of a wave and can alter its appearance in light waves. For example, polarized sunglasses remove the "glare" from sunlight by altering the oscillation pattern observed by the wearer. Choice *A*, reflection, is the bouncing back of a wave, such as in a mirror. Choice *B* is the bending of a wave as it travels from one medium to another, such as going from air to water. Choice *C*, dispersion, is the spreading of a wave through a barrier or a prism.

21. D: Radiation can be transmitted through electromagnetic waves and needs no medium to travel. Radiation can travel in a vacuum. This is how the Sun warms the Earth and typically applies to large objects with great amounts of heat, or objects that have a large difference in their heat measurements. Choice *A*, convection, involves atoms or molecules traveling from areas of high concentration to those of low concentration and they transfer energy or heat with them. Choice *B*, conduction, involves the touching or bumping of atoms or molecules in order to transfer energy or heat. Choice *C*, induction, deals with charges and does not apply to the transfer of energy or heat. Choices *A*, *B*, and *C* need a medium in which to travel, while radiation requires no medium.

22. A: 113°F

$$^0F = \frac{9}{5}(^0C) + 32$$

$$^0F = \frac{9}{5}(45) + 32$$

$$^0F = 113^0F$$

Choices *B*, *C*, and *D* all incorporate a mistake in the order of operations necessary for this calculation: divide, multiply, and then add.

23. B: A neutralization reaction produces a salt. A solid produced during a reaction is called **precipitation.** A precipitation reaction can be used for removing a salt (an ionic compound that results

from a neutralization reaction) from a solvent such as water. For water, this process is called ionization. Therefore, the products of a neutralization reaction (when an acid and base react) are a salt and water. Choice *A*, oxidation reaction, involves the transfer of an electron. Choice *C*, synthesis reaction, involves the joining of two molecules to form a single molecule. Choice *D*, decomposition reaction, involves the separation of a molecule into two other molecules.

24. B: The red-blue shift refers to the phenomenon of the Doppler effect experienced in space. An observer in space would see light approaching as having shorter wavelengths than in actuality, causing it to appear blue. When the light wave gets farther away, it would appear red, due to the apparent elongation of the wavelength. This is called the red-blue shift. The other choices are just fictional color combinations.

25. D: Circular motion occurs around an invisible line around which an object can rotate; this invisible line is called an axis. Choice *A*, center of mass, is the average placement of an object's mass. Choice *B* is not a real term. Choice *C*, elliptical, describes an elongated circle and is not a viable selection.

Arithmetic Reasoning

Problems in the **Arithmetic Reasoning** section are generally word problems, which will require the use of reasoning and mathematics to find a solution. The problems normally present some everyday situations, along with a list of possible answer choices. To succeed in this section, test takers should be comfortable solving problems involving rates, speeds, percentages, averages, fractions, and ratios. The practice problems given later will cover the different types of questions in this section, although every word problem is slightly different.

How to Prepare

These problems involve basic arithmetic skills as well as the ability to interpret a word problem to see where to apply these skills to get the answer. The basics of arithmetic and the approach to solving word problems are discussed in this section.

Note that one must practice math to become proficient. Test takers are encouraged to not just read through the material here, but also work through the practice questions and review their solutions. Just reading through examples does not necessarily mean that a student can do the problems alone; it is necessary to work through the problem and attempt to find the solution before reviewing the provided answer. Note that sometimes there can be multiple approaches to a problem, all of which can find the correct solution. What matters is getting the correct answer, so it is okay if the approach to a problem is different than the solution method described.

Basic Operations of Arithmetic

There are four basic operations used with numbers: addition, subtraction, multiplication, and division.

- **Addition** takes two numbers and combines them into a total called the sum. The **sum** is the total when combining two collections into one. If there are 5 things in one collection and 3 in another, then after combining them, there is a total of $5 + 3 = 8$. Note the order does not matter when adding numbers. For example, $3 + 5 = 8$.

- **Subtraction** is the opposite (or "inverse") operation to addition. Whereas addition combines two quantities together, subtraction takes one quantity away from another. For example, if there are 20 gallons of fuel and 5 are used, that leaves $20 - 5 = 15$ gallons remaining. Note that for subtraction, the order does matter because it makes a difference which quantity is being removed from which.

- **Multiplication** is repeated addition. 3×4 can be thought of as putting together 3 sets of items, with each set containing 4 items. The total is 12 items. Another way to think of this is to think of each number as the length of one side of a rectangle. If a rectangle is covered in tiles with 3 columns of 4 tiles each, then there are 12 tiles in total. From this, one can see that the answer is the same if the rectangle has 4 rows of 3 tiles each: $4 \times 3 = 12$. By expanding this reasoning, the order in which the numbers are multiplied does not matter.

- **Division** is the inverse of multiplication. It means taking one quantity and partitioning it into sets the size of the second quantity. If there are 16 sandwiches to be distributed evenly between 4 people, then each person gets $16 \div 4 = 4$ sandwiches. As with subtraction, the order in which the numbers appear does matter for division.

Lastly, **exponentiation** is a kind of repeated multiplication. 3^4, read "three to the fourth power," means to multiply 3 by itself 4 times: $3 \times 3 \times 3 \times 3 = 81$. The exponent 2 is the exponentiation case used most often in calculations. In that case, it is called **squaring** the number. Squaring is used in some formulas for geometric areas.

A **fraction** is a number used to express a ratio. It is written as a number x over a line with another number y underneath: $\frac{x}{y}$, and can be thought of as x out of y equal parts. The number on top (x) is called the **numerator**, and the number on the bottom (y) is called the **denominator**. It is important to remember the only restriction is that the denominator is not allowed to be 0.

Another way of thinking about fractions is that $\frac{x}{y} = x \div y$.

Two fractions can sometimes equal the same number even when they look different. The value of a fraction will remain equal when multiplying both the numerator and the denominator by the same number. The value of the fraction also does not change when dividing both the numerator and the denominator by the same number. For example, $\frac{4}{8} = \frac{2}{4} = \frac{1}{2}$. If two fractions look different, but they actually represent the same value, they are **equivalent fractions**.

A number that can divide evenly into a second number is called a **divisor** or **factor** of that second number; 3 is a divisor of 6, for example. If the numerator and denominator in a fraction have no common factors other than 1, the fraction is said to be **simplified**. $\frac{2}{4}$ is not simplified (since the numerator and denominator have a factor of 2 in common), but $\frac{1}{2}$ is simplified. When solving a problem, the final answer generally requires one to simplify the fraction.

It is often useful when working with fractions to rewrite them so they have the same denominator. This process is called finding a **common denominator**. The common denominator of two fractions needs to be a number that is a multiple of both denominators. For example, given $\frac{1}{6}$ and $\frac{5}{8}$, a common denominator is $6 \times 8 = 48$. However, there are often smaller choices for the common denominator. The smallest number that is a multiple of two numbers is called the **least common multiple** of those numbers. For this example, use the numbers 6 and 8. The multiples of 6 are 6, 12, 18, 24... and the multiples of 8 are 8, 16, 24..., so the least common multiple is 24. The two fractions are rewritten as $\frac{4}{24}, \frac{15}{24}$. Selecting the least common multiple for the denominator (called the **lowest common denominator**) is important because it will help find the solution in its simplified form.

If two fractions have a common denominator, then the numerators can be added or subtracted. For example, $\frac{4}{5} - \frac{3}{5} = \frac{4-3}{5} = \frac{1}{5}$. If the fractions are not given with the same denominator, a common denominator needs to be found before adding or subtracting them.

To multiply two fractions, the numerators are multiplied together to find the new numerator and the denominators are multiplied together to get the new denominator. For example, $\frac{3}{5} \times \frac{2}{7} = \frac{3 \times 2}{5 \times 7} = \frac{6}{35}$.

Switching the numerator and denominator is called taking the **reciprocal** of a fraction; the reciprocal of $\frac{4}{5}$ is $\frac{5}{4}$.

To divide one fraction by another, the first fraction is multiplied by the reciprocal of the second. So $\frac{3}{4} \div \frac{2}{5} = \frac{3}{4} \times \frac{5}{2} = \frac{15}{8}$.

If the numerator is smaller than the denominator, the fraction is a **proper fraction**. Otherwise, the fraction is said to be **improper**.

A **mixed number** is a number that is an integer plus some proper fraction and is written with the integer first and the proper fraction to the right of it. Any improper fraction can be rewritten as a mixed number. For instance, $\frac{5}{3} = 1\frac{2}{3}$.

Percentages are essentially just fractions out of 100 (the word comes from the Latin meaning "per one hundred") and are written with the % symbol. So, 35% means $\frac{35}{100}$. Converting from a fraction to a percent requires the fraction to be converted such that it has a denominator of 100; the percent is then the numerator. When converting from a percent to a fraction, it's important to remember that the percent is really a fraction with a denominator of 100.

To convert a decimal to a percent, the decimal point is moved two places to the right. For example, 0.89 = 89%. To convert a percent to a decimal, the decimal point is moved two places to the left. For example, 65% = 0.65.

A **ratio** compares two quantities in size and behaves much like a fraction. If a building has 10 offices and 15 employees, then the ratio of offices to employees is 10 to 15, which can also be written as 10:15. Like fractions, both numbers be can be multiplied or divided in a ratio without changing the value of the ratio. By dividing each quantity by 5, the ratio of offices to employees can also be written as 2 to 3. A ratio is usually given in the form that has the smallest possible whole numbers. As with simplifying fractions, this means the ratio is written using two numbers whose only common factor is 1.

Two quantities are in a **proportional relationship** when one quantity increases or decreases by a fixed fraction of some second quantity. Purchasing gas generally involves a proportional relationship: for each gallon of gas purchased, the price goes up by a fixed amount: Cost = Price × Quantity. All proportional relationships involve a relationship like this, where one quantity is given by multiplying the second quantity by some factor, which is called the **factor of proportionality**.

Two quantities are in an **inversely proportional** relationship when one quantity decreases as the other increases, in a relationship where the first quantity is given by the second quantity *divided* by some factor. An example of this is the time that a trip takes versus the speed travelled. This is because Time = Distance ÷ Speed. All inversely proportional problems involve a relationship of this form.

Basic Geometry Relationships

Perimeter is the measurement of a distance around something. It can be thought of as the length of the boundary, like a fence. It is found by adding together the lengths of all of the sides of a figure. Since a square has four equal sides, its perimeter can be calculated by multiplying the length of one side by 4. Thus, the formula is $P = 4 \times s$, where s equals one side. Like a square, a rectangle's perimeter is measured by adding together all of the sides. But as the sides are unequal, the formula is different. A

rectangle has equal values for its lengths (long sides) and equal values for its widths (short sides), so the perimeter formula for a rectangle is $P = l + l + w + w = 2l + 2w$, where l is length and w is width. Perimeter is measured in simple units such as inches, feet, yards, centimeters, meters, miles, etc.

In contrast to perimeter, **area** is the space occupied by a defined enclosure, like a field enclosed by a fence. It is measured in square units such as square feet or square miles. Here are some formulas for the areas of basic planar shapes:

- The area of a rectangle is $l \times w$, where w is the width and l is the length
- The area of a square is s^2, where s is the length of one side (this follows from the formula for rectangles)
- The area of a triangle with base b and height h is $\frac{1}{2}bh$
- The area of a circle with radius r is πr^2

Volume is the measurement of how much space an object occupies, like how much space is in the cube. Volume questions will typically ask how much of something is needed to completely fill the object. It is measured in cubic units, such as cubic inches. Here are some formulas for the volumes of basic three-dimensional geometric figures:

- For a regular prism whose sides are all rectangles, the volume is $l \times w \times h$, where w is the width, l is the length, and h is the height of the prism
- For a cube, which is a prism whose faces are all squares of the same size, the volume is s^3
- The volume of a sphere of radius r is given by $\frac{4}{3}\pi r^3$
- The volume of a cylinder whose base has a radius of r and which has a height of h is given by $\pi r^2 h$

Comparing Data

Comparing data sets within statistics can mean many things. The first way to compare data sets is by looking at the center and spread of each set. The center of a data set can mean two things: median or mean. The **median** is the value that's halfway into each data set when arranged in numerical order, and it splits the data into two intervals. The **mean** is the average value of the data within a set. It's calculated by adding up all of the data in the set and dividing the total by the number of data points. Outliers can significantly impact the mean. Additionally, two completely different data sets can have the same mean. For example, a data set with values ranging from 0 to 100 and a data set with values ranging from 44 to 56 can both have means of 50. The first data set has a much wider range, which is known as the **spread** of the data. This measures how varied the data is within each set.

Practice Questions

1. If a car can travel 300 miles in 4 hours, how far can it go in an hour and a half?
    a. 100 miles
    b. 112.5 miles
    c. 135.5 miles
    d. 150 miles

2. At the store, Jan spends $90 on apples and oranges. Apples cost $1 each and oranges cost $2 each. If Jan buys the same number of apples as oranges, how many oranges did she buy?
    a. 20
    b. 25
    c. 30
    d. 35

3. What is the volume of a box with rectangular sides 5 feet long, 6 feet wide, and 3 feet high?
    a. 60 cubic feet
    b. 75 cubic feet
    c. 90 cubic feet
    d. 14 cubic feet

4. A train traveling 50 miles per hour takes a trip lasting 3 hours. If a map has a scale of 1 inch per 10 miles, how many inches apart are the train's starting point and ending point on the map?
    a. 14
    b. 12
    c. 13
    d. 15

5. A traveler takes an hour to drive to a museum, spends 3 hours and 30 minutes there, and takes half an hour to drive home. What percentage of his or her time was spent driving?
    a. 15%
    b. 30%
    c. 40%
    d. 60%

6. A truck is carrying three cylindrical barrels. Their bases have a diameter of 2 feet and they have a height of 3 feet. What is the total volume of the three barrels in cubic feet?
    a. 3π
    b. 9π
    c. 12π
    d. 15π

7. Greg buys a $10 lunch with 5% sales tax. He leaves a $2 tip after his bill. How much money does he spend?
    a. $12.50
    b. $12
    c. $13
    d. $13.25

8. Marty wishes to save $150 over a 4-day period. How much must Marty save each day on average?
   a. $37.50
   b. $35
   c. $45.50
   d. $41

9. Bernard can make $80 per day. If he needs to make $300 and only works full days, how many days will this take?
   a. 6
   b. 3
   c. 5
   d. 4

10. A couple buys a house for $150,000. They sell it for $165,000. By what percentage did the house's value increase?
   a. 10%
   b. 13%
   c. 15%
   d. 17%

11. A school has 15 teachers and 20 teaching assistants. They have 200 students. What is the ratio of faculty to students?
   a. 3:20
   b. 4:17
   c. 5:54
   d. 7:40

12. A map has a scale of 1 inch per 5 miles. A car can travel 60 miles per hour. If the distance from the start to the destination is 3 inches on the map, how long will it take the car to make the trip?
   a. 12 minutes
   b. 15 minutes
   c. 17 minutes
   d. 20 minutes

13. Taylor works two jobs. The first pays $20,000 per year. The second pays $10,000 per year. She donates 15% of her income to charity. How much does she donate each year?
   a. $4500
   b. $5000
   c. $5500
   d. $6000

14. A box with rectangular sides is 24 inches wide, 18 inches deep, and 12 inches high. What is the volume of the box in cubic feet?
   a. 2
   b. 3
   c. 4
   d. 5

15. Kristen purchases $100 worth of CDs and DVDs. The CDs cost $10 each and the DVDs cost $15. If she bought four DVDs, how many CDs did she buy?
a. 5
b. 6
c. 3
d. 4

16. If Sarah reads at an average rate of 21 pages in four nights, how long will it take her to read 140 pages?
a. 6 nights
b. 26 nights
c. 8 nights
d. 27 nights

17. Mom's car drove 72 miles in 90 minutes. There are 5280 feet per mile. How fast did she drive in feet per second?
a. 0.8 feet per second
b. 48.9 feet per second
c. 0.009 feet per second
d. 70.4 feet per second

18. This chart indicates how many sales of CDs, vinyl records, and MP3 downloads occurred over the last year. Approximately what percentage of the total sales was from CDs?

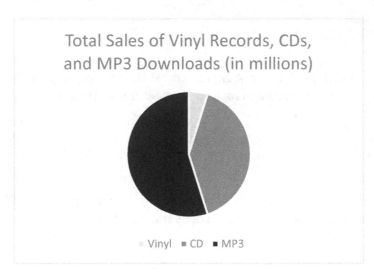

Total Sales of Vinyl Records, CDs, and MP3 Downloads (in millions)

Vinyl ■ CD ■ MP3

a. 55%
b. 25%
c. 40%
d. 5%

19. After a 20% sale discount, Frank purchased a new refrigerator for $850. How much did he save from the original price?
    a. $170
    b. $212.50
    c. $105.75
    d. $200

20. Alan currently weighs 200 pounds, but he wants to lose weight to get down to 175 pounds. What is this difference in kilograms? (1 pound is approximately equal to 0.45 kilograms.)
    a. 9 kg
    b. 11.25 kg
    c. 78.75 kg
    d. 90 kg

21. A student gets an 85% on a test with 20 questions. How many questions did the student solve correctly?
    a. 15
    b. 16
    c. 17
    d. 18

22. Johnny earns $2334.50 from his job each month. He pays $1437 for monthly expenses. Johnny is planning a vacation in 3 months' time that he estimates will cost $1750 total. How much will Johnny have left over from three months' of saving after he pays for his vacation?
    a. $948.50
    b. $584.50
    c. $852.50
    d. $942.50

23. Dwayne has received the following scores on his math tests: 78, 92, 83, 97. What score must Dwayne get on his next math test to have an overall average of 90?
    a. 89
    b. 98
    c. 95
    d. 100

24. What is the overall median of Dwayne's current scores: 78, 92, 83, 97?
    a. 80.5
    b. 85
    c. 87.5
    d. 83

25. In Jim's school, there are 3 girls for every 2 boys. There are 650 students in total. Using this information, how many students are girls?
    a. 260
    b. 130
    c. 65
    d. 390

26. Kimberley earns $10 an hour babysitting, and after 10 p.m., she earns $12 an hour, with the amount paid being rounded to the nearest hour accordingly. On her last job, she worked from 5:30 p.m. to 11 p.m. In total, how much did Kimberley earn for that job?
    a. $45
    b. $57
    c. $62
    d. $42

27. Keith's bakery had 252 customers go through its doors last week. This week, that number increased to 378. By what percentage did his customer volume increase?
    a. 26%
    b. 50%
    c. 35%
    d. 12%

28. A family purchases a vehicle in 2005 for $20,000. In 2010, they decide to sell it for a newer model. They are able to sell the car for $8,000. By what percentage did the value of the family's car drop?
    a. 40%
    b. 68%
    c. 60%
    d. 33%

29. In May of 2010, a couple purchased a house for $100,000. In September of 2016, the couple sold the house for $93,000 so they could purchase a bigger one to start a family. How many months did they own the house?
    a. 76
    b. 54
    c. 85
    d. 93

30. At the beginning of the day, Xavier has 20 apples. At lunch, he meets his sister Emma and gives her half of his apples. After lunch, he stops by his neighbor Jim's house and gives him 6 of his apples. He then uses ¾ of his remaining apples to make an apple pie for dessert at dinner. At the end of the day, how many apples does Xavier have left?
    a. 4
    b. 6
    c. 2
    d. 1

Answer Explanations

1. B: 300 miles in 4 hours is 300/4 = 75 miles per hour. In 1.5 hours, the car will go 1.5 × 75 miles, or 112.5 miles.

2. C: One apple/orange pair costs $3 total. Therefore, Jan bought 90/3 = 30 total pairs, and hence, she bought 30 oranges.

3. C: The formula for the volume of a box with rectangular sides is the length times width times height, so $5 \times 6 \times 3 = 90$ cubic feet.

4. D: First, the train's journey in the real word is 3 x 50 = 150 miles. On the map, 1 inch corresponds to 10 miles, so there is 150/10 = 15 inches on the map.

5. B: The total trip time is 1 + 3.5 + 0.5 = 5 hours. The total time driving is 1 + 0.5 = 1.5 hours. So, the fraction of time spent driving is 1.5/5 or 3/10. To get the percentage, convert this to a fraction out of 100. The numerator and denominator are multiplied by 10, with a result of 30/100. The percentage is the numerator in a fraction out of 100, so 30%.

6. B: The formula for the volume of a cylinder is $\pi r^2 h$, where r is the radius and h is the height. The diameter is twice the radius, so these barrels have a radius of 1 foot. That means each barrel has a volume of $\pi \times 1^2 \times 3 = 3\pi$ cubic feet. Since there are three of them, the total is $3 \times 3\pi = 9\pi$ cubic feet.

7. A: The tip is not taxed, so he pays 5% tax only on the $10. 5% of $10 is $0.05 \times 10 = \$0.50$. Add up $10 + $2 + $0.50 to get $12.50.

8. A: The first step is to divide up $150 into four equal parts. 150/4 is 37.5, so she needs to save an average of $37.50 per day.

9. D: 300/80 =30/8 = 15/4 =3.75. But Bernard is only working full days, so he will need to work 4 days, since 3 days are not sufficient.

10. A: The value went up by $165,000 − $150,000 = $15,000. Out of $150,000, this is $\frac{15,000}{150,000} = \frac{1}{10}$. Convert this to having a denominator of 100, the result is $\frac{10}{100}$ or 10%.

11. D: The total faculty is 15 + 20 = 35. Therefore, the faculty to student ratio is 35:200. Then, to simplify this ratio, both the numerator and the denominator are divided by 5, since 5 is a common factor of both, which yields 7:40.

12. B: The journey will be $5 \times 3 = 15$ miles. A car travelling at 60 miles per hour is travelling at 1 mile per minute. So, it will take 15/1 = 15 minutes to take the journey.

13. A: Taylor's total income is $20,000 + $10,000 = $30,000. 15% of this is $\frac{15}{100} = \frac{3}{20}$. So:

$$\frac{3}{20} \times \$30,000 = \frac{90,000}{20}$$

$$\frac{9000}{2} = \$4500$$

14. B: Since the answer will be in cubic feet rather than inches, the first step is to convert from inches to feet for the dimensions of the box. There are 12 inches per foot, so the box is 24/12 = 2 feet wide, 18/12 = 1.5 feet deep, and 12/12 = 1 foot high. The volume is the product of these three together:

$$2 \times 1.5 \times 1 = 3 \text{ cubic feet}$$

15. D: Kristen bought four DVDs, which would cost a total of $4 \times 15 = \$60$. She spent a total of $100, so she spent $100 – $60 = $40 on CDs. Since they cost $10 each, she must have purchased 40/10 = four CDs.

16. D: This problem can be solved by setting up a proportion involving the given information and the unknown value. The proportion is:

$$\frac{21 \, pages}{4 \, nights} = \frac{140 \, pages}{x \, nights}$$

Solving the proportion by cross-multiplying, the equation becomes $21x = 4 \times 140$, where $x = 26.67$. Since it is not an exact number of nights, the answer is rounded up to 27 nights. Twenty-six nights would not give Sarah enough time.

17. D: This problem can be solved by using unit conversion. The initial units are miles per minute. The final units need to be feet per second. Converting miles to feet uses the equivalence statement 1 mile = 5,280 feet. Converting minutes to seconds uses the equivalence statement 1 minute = 60 seconds. Setting up the ratios to convert the units is shown in the following equation:

$$\frac{72 \, miles}{90 \, minutes} * \frac{1 \, minute}{60 \, seconds} * \frac{5280 \, feet}{1 \, mile} = 70.4 \text{ feet per second}$$

The initial units cancel out, and the new units are left.

18. C: The sum total percentage of a pie chart must equal 100%. Since the CD sales take up less than half of the chart and more than a quarter (25%), it can be determined to be 40% overall. This can also be measured with a protractor. The angle of a circle is 360°. Since 25% of 360 would be 90° and 50% would be 180°, the angle percentage of CD sales falls in between; therefore, it would be Choice *C*.

19. B: Since $850 is the price *after* a 20% discount, $850 represents 80% of the original price. To determine the original price, set up a proportion with the ratio of the sale price (850) to original price (unknown) equal to the ratio of sale percentage:

$$\frac{850}{x} = \frac{80}{100}$$

(where *x* represents the unknown original price)

To solve a proportion, cross multiply the numerators and denominators and set the products equal to each other: $(850) \times (100) = (80) \times (x)$

Multiplying each side results in the equation 85,000 = 80x.

To solve for x, both sides get divided by 80: $\frac{85,000}{80} = \frac{80x}{80}$, resulting in $x = 1062.5$. Remember that x represents the original price. Subtracting the sale price from the original price ($1062.50 – $850) indicates that Frank saved $212.50.

20. B: Using the conversion rate, the projected weight loss of 25 pounds is multiplied by 0.45 $\frac{kg}{lb}$ to get the amount in kilograms (11.25 kg).

21. C: 85% of a number means that number should be multiplied by 0.85: $0.85 \times 20 = \frac{85}{100} \times \frac{20}{1}$, which can be simplified to $\frac{17}{20} \times \frac{20}{1} = 17$. The answer is C.

22. D: The first step is to subtract $1437 from $2334.50 to find Johnny's monthly savings; this equals $897.50. Then, this amount is multiplied by 3 to find out how much he will have after three months before he pays for his vacation: this equals $2692.50. Finally, the cost of the vacation ($1750) is subtracted from this amount to find how much Johnny will have left: $942.50.

23. D: To find the average of a set of values, the values are added together and then this sum is divided by the total number of values. In this case, the unknown value of what Dwayne needs to score on his next test needs to be added, in order to solve it.

$$\frac{78 + 92 + 83 + 97 + x}{5} = 90$$

The unknown value is added to the new average total, which is 5. Then, each side is multiplied by 5 to simplify the equation, resulting in:

$$78 + 92 + 83 + 97 + x = 450$$

$$350 + x = 450$$

$$x = 100$$

Dwayne would need to get a perfect score of 100 in order to get an average of at least 90.

This answer can be confirmed by substituting it back into the original formula.

$$\frac{78 + 92 + 83 + 97 + 100}{5} = 90$$

24. C: For an even number of total values, the median is calculated by finding the mean or average of the two middle values once all values have been arranged in ascending order from least to greatest. In this case, $(92 + 83) \div 2$ would equal the median 87.5, Choice C.

25. D: Three girls for every two boys can be expressed as a ratio: 3:2. This can be visualized as splitting the school into 5 groups: 3 girl groups and 2 boy groups. The number of students that are in each group can be found by dividing the total number of students by 5:

650 divided by 5 equals 1 part, or 130 students per group

To find the total number of girls, the number of students per group (130) is multiplied by how the number of girl groups in the school (3). This equals 390, Choice *D*.

26. C: Kimberley worked 4.5 hours at the rate of $10/h and 1 hour at the rate of $12/h. The problem states that her pay is rounded to the nearest hour, so the 4.5 hours would round up to 5 hours at the rate of $10/h.

$$(5h) \times \left(^{\$10}/_h\right) + (1h) \times \left(^{\$12}/_h\right) = \$50 + \$12 = \$62$$

27. B: The first step is to calculate the difference between the larger value and the smaller value.

$$378 - 252 = 126$$

To calculate this difference as a percentage of the original value, and thus calculate the percentage *increase*, 126 is divided by 252, then this result is multiplied by 100 to find the percentage = 50%, answer *B*.

28. C: In order to find the percentage by which the value of the car has been reduced, the current cash value should be subtracted from the initial value and then the difference divided by the initial value. The result should be multiplied by 100 to find the percentage decrease.

$$\frac{20,000 - 8,000}{20,000} = .6$$

$$(.60) \times 100 = 60\%$$

29. A: This problem can be solved by simple multiplication and addition. Since the sale date is over six years apart, 6 can be multiplied by 12 for the number of months in a year, and then the remaining 4 months can be added.

$$(6 \times 12) + 4 = ?$$

$$72 + 4 = 76$$

30. D: This problem can be solved using basic arithmetic. Xavier starts with 20 apples, then gives his sister half, so 20 divided by 2.

$$\frac{20}{2} = 10$$

He then gives his neighbor 6, so 6 is subtracted from 10.

$$10 - 6 = 4$$

Lastly, he uses ¾ of his apples to make an apple pie, so to find remaining apples, the first step is to subtract ¾ from one and then multiply the difference by 4.

$$\left(1 - \frac{3}{4}\right) \times 4 = ?$$

$$\left(\frac{4}{4} - \frac{3}{4}\right) \times 4 = ?$$

$$\left(\frac{1}{4}\right) \times 4 = 1$$

Word Knowledge

Defining Words and English Origins

A **word** is a group of letters joined to form a single meaning. On their own, letters represent single sounds, but when placed together in a certain order, letters represent a specific image in the reader's mind in a way that provides meaning. Words can be *nouns, verbs, adjectives,* and *adverbs,* among others. Words also represent a verb tense of *past, present,* or *future.* Words allow for effective communication for commerce, social progress, technical advances, and much more. Simply put, words allow people to understand one another and create meaning in a complex world.

Throughout history, English words were shaped by other cultures and languages, such as Greek, Latin, French, Spanish, German, and others. They were borne from inventions, discoveries, and literary works, such as plays or science fiction novels. Others formed by shortening words that were already in existence. Some words evolved from the use of acronyms, such as *radar* (Radio Detection and Ranging) and *scuba* (Self-Contained Underwater Breathing Apparatus). The English language will continue to evolve as the needs and values of its speakers evolve.

Word Formation

How do English words form? They can be single root words, such as *love, hate, boy,* or *girl.* A **root word** is a word in its most basic form that carries a clear and distinct meaning. Complex English words combine affixes with root words. Some words have no root word, but are instead formed by combining various affixes, such as reject: *re-* is defined as repeating an action or actions, and *-ject* means throw or thrown. Therefore, **reject** is defined as the act of being thrown back. Words consisting of affixes alone are not the norm. Most words consist of either root words on their own or with the addition of affixes. To **affix** is to attach to something. Therefore, affixes in linguistic study are groups of letters that attach themselves to the beginning, middle, or end of root words to enhance or alter the word's meaning. Affixes added to verbs can change the word's tense, and affixes added to nouns can change the word's part of speech from noun to adjective, verb, or adverb.

Roots and Root Words

Words that exist on their own, without affixes, are root words. **Root words** are words written in their most basic form, and they carry a clear and distinct meaning. Consider the word *safe.* The root word, *safe,* acts as both a noun and adjective, and stands on its own, carrying a clear and distinct meaning.

The root of a word however, is not necessarily a part of the word that can stand on its own, although it does carry meaning. Since many English words come from Latin and Greek roots, it's helpful to have a

general understanding of roots. Here is a list of common Greek and Latin roots used in the English language:

Root	Definition	Example
ami	love	amiable
ethno	race	ethnological
infra	beneath or below	infrastructure
lun	moon	Lunar
max	greatest	maximum
pent/penta	five	pentagon
sol	sun	Solar
vac	empty	Vacant

Affixes

Affixes are groups of letters that when added to the beginning or ending of root words, or are attachments within a root or root word itself, can:

- Intensify the word's meaning
- Create a new meaning
- Somewhat alter the existing meaning
- Change the verb tense
- Change the part of speech

There are three types of affixes: prefixes, suffixes, and infixes.

Prefixes

Prefixes are groups of letters attached to the beginning of a root word. *Pre-* refers to coming before, and *fix* refers to attaching to something. Consider the example of the root word *freeze*:

- Freeze: verb – to change from a liquid to solid by lowering the temperature to a freezing state.

- *Anti*freeze: noun – a liquid substance that prevents freezing when added to water, as in a vehicle's radiator.

By adding the prefix *anti-* to the root word *freeze*, the part of speech changed from verb to noun, and completely altered the meaning. *Anti-* as a prefix always creates the opposite in meaning, or the word's antonym.

By having a basic understanding of how prefixes work and what their functions are in a word's meaning, English speakers strengthen their fluency. Here is a list of some common prefixes in the English language, accompanied by their meanings:

Prefix	Definition	Example
ante-	before	antecedent
ex-	out/from	expel
inter-	between/among	intergalactic
multi-	much/many	multitude
post-	after	postscript
sub-	under	submarine
trans-	move between/across	transport
uni-	single/one	universe

Suffixes

Suffixes are groups of letters attached to the ending of a root or root word. Like prefixes, suffixes can:

- Intensify the word's meaning
- Create a new meaning
- Somewhat alter the existing meaning
- Change the verb tense
- Change the part of speech

Consider the example of the root word *fish* when suffixes are added:

- Fish: noun – a cold-blooded animal that lives completely in water and possesses fins and gills.
- Fishing: noun – I love the sport of fishing.
- Fishing: verb – Are you fishing today?

With the addition of the suffix *-ing*, the meaning of root word *fish* is altered, as is the part of speech.

A verb tense shift is made with the addition of the suffix *-ed*:

- Jump: present tense of to jump as in "I jump."
- Jumped: past tense of to jump as in "I jumped."
- Climb: present tense of to climb as in "We climb."
- Climbed: past tense of to climb as in "We climbed."

Here are a few common suffixes in the English language, along with their meanings:

Suffix	Meaning	Example
-ed	past tense	cooked
-ing	materials, present action	clothing
-ly	in a specific manner	lovely
-ness	a state or quality	brightness
-ment	action	enjoyment
-script	to write	transcript
-ee	receiver/performer	nominee
-ation/-ion	action or process	obligation

Infixes

Infixes are letters that attach themselves inside the root or root words. They generally appear in the middle of the word and are rare in the English language. Easily recognizable infixes include parents-in-law, passers-by, or cupsful. Notice the -*s* is added inside the root word, making the word plural.

Special types of infixes, called **tmesis** words, are made from inserting an existing word into the middle of another word or between a compound word, creating a new word. Tmesis words are generally used in casual dialogue and slang speech. They add emphasis to the word's overall meaning and to evoke emotion on the part of the reader. Examples are fan-*bloody*-tastic and un-*freaking*-believable.

Tmesis words have been in existence since Shakespeare's time, as in this phrase from *Romeo and Juliet*, "...he is some *other*where." Shakespeare split up the compound word *somewhere* by inserting the word *other* between the two root words.

Compound Words

A **compound word** is created with the combination of two shorter words. To be a true compound word, two shorter words are combined, and the meaning of the longer word retains the meaning of the two shorter words. Compound words enhance the overall meaning, giving a broader description. There are three types of compound words: closed, hyphenated, and open.

Closed Compound Words

There was a time when closed compound words were not considered legitimate words. Over time and with continued, persistent use, they found a place in the English language. A **closed compound word**

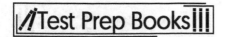

refers to a word that shows no separation between the two shorter words from which it is composed. Some examples of closed compound words are:

Closed Compound Word	Individual Words	Meaning
bookshelf	book/shelf	a shelf that holds books
doorstop	door/stop	an object to hold a door open
bedroom	bed/room	a room where one sleeps
bathroom	bath/room	a room where one bathes
backyard	back/yard	a yard in the back of a building
nightstand	night/stand	a small table beside one's bed

Hyphenated Compound Words

As the name suggests, **hyphenated compound words** include a hyphen that separates the two shorter words within the longer word. Some examples of hyphenated compound words are:

Hyphenated Compound Words	Individual Words	Meaning
self-service	self/serve	the act of serving one's self
color-blind	color/blind	incapable of accurately distinguishing colors
check-in	check/in	the act of registering as in attendance
year-round	year/round	any affair that takes place throughout the year
toll-free	toll/free	no application of toll/no charge
sugar-coated	sugar/coated	anything sweetened or coated with sugar

Open Compound Words

Open compound words appear as individual words but are they dependent on their partners to form the complete meaning of the compound word. Individually, the words may have meanings that are different than that of the pair together (the open compound word). For example, in *real estate, real* and *estate* have their own meanings that are different than the unique meaning of the compound word. Open compound words are separated from each other by a single space and do not require a hyphen. Some examples of open compound words are:

Open Compound Words	Meaning
polka dot	a repeated circular dot that forms a pattern
sleeping bag	a special bag used to sleep in (usually when camping)
solar system	the collection of planets that orbit around the sun
tape recorder	magnetic tape used to record sound
middle class	social group that is considered above lower class and below upper class
family room	a specific room for relaxation and entertainment used by all family members

Parts of Speech

Also referred to as word classes, **parts of speech** refer to the various categories in which words are placed. Words can be placed in any one or a combination of the following categories:

- Nouns
- Determiners
- Pronouns
- Verbs
- Adjectives
- Adverbs
- Prepositions
- Conjunctions

Understanding the various parts of speech used in the English language helps readers to better understand the written language.

Nouns

A **noun** is defined as any word that represents a person, place, animal, object, or idea. Nouns can identify a person's title or name, a person's gender, and a person's nationality, such as banker, commander-in-chief, female, male, or an American.

With animals, nouns identify the kingdom, phylum, class, etc. For example: the animal is an *elephant*, the phylum is *chordata*, and the class is *mammalia*. It should be noted that the words *animal, phylum,* and *class* in the previous sentence are also nouns.

When identifying places, nouns refer to a physical location, a general vicinity, or the proper name of a city, state, or country. Some examples include the *desert*, the *East*, *Phoenix*, the *bathroom*, *Arizona*, an *office*, or the *United States*.

There are eight types of nouns: common, proper, countable, uncountable, concrete, abstract, compound, and collective.

Common nouns are used in general terms, without specific identification. Examples include *girl, boy, country*, or *school*. Proper nouns refer to the specific proper name given to people, places, animals, or entities, such as *Melissa, Martin, Italy*, or *Harvard*.

Countable nouns can be counted: *one car, two cars*, or *three cars*. **Uncountable nouns** cannot be counted, such as *air, liquid*, or *gas*.

To be abstract is to exist, but only in thought or as an idea. An **abstract noun** cannot be physically touched, seen, smelled, heard, or tasted. Examples include *chivalry, day, fear, thought, truth, friendship*, or *freedom*.

To be **concrete** is to be seen, touched, tasted, heard, and/or smelled. Examples include *pie, snow, tree, bird, desk, hair*, or *dog*.

A **compound noun** is another term for an open compound word. Any noun that is written as two nouns that together form a specific meaning is a compound noun. For example, *post office, ice cream*, or *swimming pool*.

A **collective noun** refers to groups or collection of things that together form the whole. The members of the group are often people or individuals. Examples include *orchestra, squad, committee*, or the *majority*. It should be noted that the nouns in these examples are singular, but the word itself refers to a group containing more than one individual.

Determiners

Determiners modify a noun and usually refer to something specific. Determiners fall into one of four categories: *articles, demonstratives, quantifiers*, or *possessive determiners*.

Articles can be both definite articles, as in *the*, and indefinite as in *a, an, some*, or *any*:

> *The* man with *the* red hat.

> *A* flower growing in *the* yard.

> *Any* person who visits *the* store.

There are four different types of demonstratives: *this, that, these*, and *those*.

True demonstrative words will not directly precede the noun of the sentence but will be the noun. Some examples:

> *This* is the one.

> *That* is the place.

> *Those* are the files.

Once a demonstrative is placed directly in front of the noun, it becomes a demonstrative pronoun:

> *This* one is perfect.

> *That* place is lovely.

> *Those* boys are annoying.

Quantifiers proceed nouns to give additional information to the noun about how much or how many is referred to. They can be used with countable and uncountable nouns:

> She bought *plenty* of apples.

> *Few* visitors came.

> I got a *little* change.

Possessive determiners, sometimes called **possessive adjectives**, indicate possession. They are the possessive forms of personal pronouns, such as *for my, your, hers, his, its, their,* or *our*:

> That is *my* car.

> Tom rode *his* bike today.

> Those papers are *hers.*

Pronouns

Pronouns are words that stand in place of nouns. There are three different types of pronouns: **subjective pronouns** (*I, you, he, she, it, we, they*), **objective pronouns** (*me, you, him, her it, us, them*), and **possessive pronouns** (*mine, yours, his, hers, ours, theirs*).

Note that some words are found in more than one pronoun category. See examples and clarifications are below:

> *You* are going to the movies.

In the previous sentence, *you* is a subjective pronoun; it is the subject of the sentence and is performing the action.

> I threw the ball to *you.*

Here, *you* is an objective pronoun; it is receiving the action and is the object of the sentence.

> We saw *her* at the movies.

Her is an objective pronoun; it is receiving the action and is the object of the sentence.

> The house across the street from the park is *hers.*

In this example, *hers* is a possessive pronoun; it shows possession of the house.

Verbs

Verbs are words in a sentence that show action or state. A sentence must contain a subject and a verb. Without a verb, a sentence is incomplete. Verbs may be in the present, past, or future tenses. To form auxiliary, or helping, verbs are required for some tenses in the future and in the past.

> I *see* the neighbors across the street.

See is an action.

> We *were eating* at the picnic.

Eating is the main action, and the verb *were* is the past tense of the verb *to be*, and is the helping or auxiliary verb that places the sentence in the past tense.

> You *will turn* 20 years of age next month.

Turn is the main verb, but *will* is the helping verb to show future tense of the verb *to be*.

Adjectives

Adjectives are a special group of words used to modify or describe a noun. Adjectives provide more information about the noun they modify. For example:

> The boy went to school. (There is no adjective.)

Rewriting the sentence, adding an adjective to further describe the boy and/or the school yields:

> The *young* boy went to the *old* school. (The adjective *young* describes the boy, and the adjective *old* describes the school.)

Adverbs

Adverb can play one of two roles: to modify the adjective or to modify the verb. For example:

> The young boy went to the old school.

We can further describe the adjectives *young* and *old* with adverbs placed directly in front of the adjectives:

> The *very* young boy went to the *very* old school. (The adverb *very* further describes the adjectives *young* and *old*.)

Other examples of using adverbs to further describe verbs include:

> The boy *slowly* went to school.

> The boy *happily* went to school.

The adverbs *slowly* and *happily* further modify the verbs.

Prepositions

Prepositions are special words that generally precede a noun. Prepositions clarify the relationship between the subject and another word or element in the sentence. They clarify time, place, and the positioning of the subjects and objects in a sentence. Common prepositions in the English language

include: *near, far, under, over, on, in, between, beside, of, at, until, behind, across, after, before, for, from, to, by,* and *with*.

Conjunctions

Conjunctions are a group of unique words that connect clauses or sentences. They also work to coordinate words in the same clause. It is important to choose an appropriate conjunction based on the meaning of the sentence. Consider these sentences:

> I really like the flowers, *however* the smell is atrocious.

> I really like the flowers, *besides* the smell is atrocious.

The conjunctions *however* and *besides* act as conjunctions, connecting the two ideas: *I really like the flowers,* and, *the smell is atrocious*. In the second sentence, the conjunction *besides* makes no sense and would confuse the reader. Conjunctions must be chosen that clearly state the intended message without ambiguity.

Some conjunctions introduce an opposing opinion, thought, or fact. They can also reinforce an opinion, introduce an explanation, reinforce cause and effect, or indicate time. For example:

> **Opposition:** She wishes to go to the movies, *but* she doesn't have the money.

> **Cause and effect:** The professor became ill, *so* the class was postponed.

> **Time:** They visited Europe *before* winter came.

Each conjunction serves a specific purpose in uniting two separate ideas. Below are common conjunctions in the English language:

Opposition	Cause & Effect	Reinforcement	Time	Explanation
however	Therefore	besides	afterward	for example
nevertheless	as a result	anyway	before	in other words
but	because of this	after all	firstly	for instance
although	Consequently	furthermore	next	such as

Synonyms

Synonyms are words that mean the same or nearly the same thing in the same language. When presented with several words and asked to choose the synonym, more than one word may be similar to the original. However, one word is generally the strongest match. Synonyms should always share the same part of speech. For instance, *shy* and *timid* are both adjectives and hold similar meanings. The words *shy* and *loner* are similar, but *shy* is an adjective while *loner* is a noun. Another way to test for the best synonym is to reread the sentence with each possible word and determine which one makes the most sense. Consider the following sentence: *He will love you forever.*

Now consider the words: *adore, sweet, kind*, and *nice*. They seem similar, but when used in the following applications with the initial sentence, not all of them are synonyms for *love*.

He will *adore* you forever.

He will *sweet* you forever.

He will *kind* you forever.

He will *nice* you forever.

In the first sentence, the word *love* is used as a verb. The best synonym from the list that shares the same part of speech is *adore*. *Adore* is a verb, and when substituted in the sentence, it is the only substitution that makes grammatical and semantic sense.

Synonyms can be found for nouns, adjectives, verbs, adverbs, and prepositions. Here are some examples of synonyms from different parts of speech:

- **Nouns**: clothes, wardrobe, attire, apparel
- **Verbs**: run, sprint, dash
- **Adjectives**: fast, quick, rapid, swift
- **Adverbs**: slowly, nonchalantly, leisurely
- **Prepositions**: near, proximal, neighboring, close

Here are several more examples of synonyms in the English language:

Word	Synonym	Meaning
smart	intelligent	having or showing a high level of intelligence
exact	specific	clearly identified
almost	nearly	not quite but very close
to annoy	to bother	to irritate
to answer	to reply	to form a written or verbal response
building	edifice	a structure that stands on its own with a roof and four walls
business	commerce	the act of purchasing, negotiating, trading, and selling
defective	faulty	when a device is not working or not working well

Antonyms

Antonyms are words that are complete opposites. As with synonyms, there may be several words that represent the opposite meaning of the word in question. When choosing an antonym, one should

choose the word that represents as close to the exact opposite in meaning as the given word, and ensure it shares the same part of speech. Here are some examples of antonyms:

- Nouns: predator – prey
- Verbs: love – hate
- Adjectives: good – bad
- Adverbs: slowly – swiftly
- Prepositions: above – below

Homonyms

Homonyms are words that sound alike but carry different meanings. There are two different types of homonyms: homophones and homographs.

Homophones

Homophones are words that sound alike but carry different meanings and spellings. In the English language, there are several examples of homophones. Consider the following list:

Word	Meaning	Homophone	Meaning
I'll	I + will	aisle	a specific lane between seats
allowed	past tense of the verb, 'to allow'	aloud	to utter a sound out loud
eye	a part of the body used for seeing	I	first-person singular
ate	the past tense of the verb, 'to eat'	eight	the number preceding the number nine
peace	the opposite of war	piece	part of a whole
seas	large bodies of natural water	seize	to take ahold of/to capture

Homographs

Homographs are words that share the same spelling but carry different meanings and different pronunciations. Consider the following list:

Word	Meaning	Homograph	Meaning
bass	fish	bass	musical instrument
bow	a weapon used to fire arrows	bow	to bend
Polish	of or from Poland	polish	a type of shine (n); to shine (v)
desert	dry, arid land	desert	to abandon

Types of Sentences

The English language constructs three types of sentences: simple sentences, compound sentences, and complex sentences.

91

Simple Sentences

Simple sentences are sentences where one idea or action is expressed. Simple sentences can contain conjunctions joining two or more subjects, or two or more objects. However, a simple sentence contains only one verb or action. There is only one complete thought.

In the examples below, there is a simple subject, a simple object, and one verb.

> I see the cat.

> The cat sees me.

> The cat doesn't see me.

In contrast, the use of the conjunction *and* creates a compound subject:

> *Mary and I* see the cat.

Mary and I acts as the compound subject—two or more subjects connected by the conjunction *and*. This is still a simple sentence that has one compound subject, one verb, and one complete thought.

Below is an example of a simple sentence with a compound object joined by the conjunction *and*.

> The cat sees *Mary and me.*

While there is a compound object, it is a simple sentence because it contains one subject, one verb, and forms a complete thought.

Compound sentences

Compound sentences are sentences where two **independent clauses** (complete sentences that can stand on their own) are connected with a coordinating conjunction. For example:

> I see the cat, *and* the cat sees me.

> I see the cat, *but* the cat doesn't see me.

The independent clauses can be broken apart and written independently as simple sentences:

> I see the cat. The cat sees me.

> I see the cat. The cat doesn't see me.

Complex Sentences

When a sentence contains one independent clause and one or more dependent clauses, it is a **complex sentence**. A dependent clause is a part of a sentence that will not make semantic or grammatical sense on its own. For example:

> I watched the cat while it slept.

The first part of sentence has an independent clause that can stand on its own: *I watched the cat*.

The second part of the sentence contains a dependent clause that, on its own, is nonsensical and an incomplete sentence: *While it slept.*

It is important to choose the appropriate conjunction for a given sentence, as each conjunction holds a distinct meaning. Consider the following sentences and determine which makes more sense:

> I watched the cat *although* it slept.

> I watched the cat *while* it slept.

The second sentence employs the more appropriate conjunction. The use of the conjunction, *although*, in the first sentence, creates a nonsensical sentence.

Literary Devices

Literary devices are special writing techniques used to create a specific effect. Writers and poets use literary devices that may cultivate anger, amusement, compassion, sadness, fear, or excitement. Literary devices are also used to reinforce a feeling or an idea, or to help the reader clearly see the images being described. Below are some examples of literary devices:

Personification
When an inanimate object is given human or natural qualities, the literary device used is referred to as **personification**. Consider the following poem written by an unknown author. See how the word *love* is personified, or brought to life:

> Your love tenderly rests upon my lips

> And whispers sweet words in my ear

> It gently caresses my cheek

> And calms my every fear.

> Your love tenderly rests upon my lips

> And shines with radiant light

> It guides me through the darkness

> And holds me all through the night

Since love cannot literally rest, whisper, caress, guide, or hold anyone or anything, the poet has employed personification to create a sense of endearment.

Onomatopoetic Words
The words *drip, crack, bang, buzz, crunch, zip, boom,* and *splash* are examples of **onomatopoetic** words. A finite list, onomatopoetic words create the sound associated with the word itself. In the sentences: *The bee buzzed right on by,* and *Drip, drip, drip went the tap,* we can "hear" the sounds the bee and the tap make by employing the literary device known as onomatopoeia.

Similes Versus Metaphors
Often used to show comparisons between two different objects, opinions, or ideas, similes and metaphors are frequently found in poetry and other literary works.

Similes show a comparison between two different things using the words, *like* or *as*. For example:

His smile was *as* bright *as* the stars in the sky.

Life is *like* a box of chocolates.

In the first example, *his smile* is compared to the brightness of the stars, to convey the message that it is a wide, brilliant smile. In the second example, readers get the impression that life is full of pleasant (or not so pleasant) surprises, just like *a box of chocolates*.

Metaphors compare two seemingly different things that share similar characteristics, that are perhaps hidden. Unlike similes, they do not use comparison words. Instead, they directly equate the two entities, often by simply using the word *is*. The following famous metaphors conjure up powerful images; it was likely the writers' intentions to instill a solemn, reflective state in the reader.

Dying *is* a wild night and a new road. – Emily Dickenson

Conscience *is* a man's compass. – Vincent Van Gogh

Hyperboles Versus Understatements

Hyperboles are used in literature and common dialogue with the sole intention of making a point. Sometimes humorous and sometimes serious, hyperboles are exaggerations of the truth and can be an effective way to convey a message. For example:

I'm so hungry, I could eat a horse.

He's older than the hills.

I'm dying of thirst.

She has a million things to do.

Understatements downplay something, someone, or an event. Understatements can instill humor or convey a clear message. For example, after a major car wreck, one might say: *There is a little scratch on the side of the car.* As another example, when discussing a violent rainstorm, someone may say: *A little rain is trickling down.*

Literal Versus Figurative Language

When language is written in its most basic sense, using the exact language to convey a direct meaning, it is said to be **literal**. For example:

The man is sleeping on the chair.

He closed the front door.

Figurative language is the opposite of literal. Instead of a sentence written in its most basic sense, it is written using a figure of speech that is not meant to be taken literally or using the direct meaning of the

words themselves. Metaphors, similes, hyperboles, personification, and understatements are all examples of figurative language. For example:

You are a couch potato.

America is a melting pot.

You are my sunshine.

Practice Questions

Below are some practice questions similar to those that will be encountered on the Word Knowledge portion of the exam. Match the word you are given with the answer choice that best matches its meaning.

1. "I promise, I did not **instigate** the fight."
   a. begin
   b. ponder
   c. swim
   d. overwhelm

2. **Fortify** most nearly means
   a. foster
   b. barricade
   c. strengthen
   d. undermine

3. My girlfriend said, "Either you marry me, or I'm leaving you!" Not a very pleasant **ultimatum**.
   a. invitation
   b. journey
   c. greeting
   d. threat

4. **Wave** most nearly means
   a. flourish
   b. sink
   c. tide
   d. stagnant

5. Space is often referred to as the great **void**.
   a. around
   b. age
   c. empty
   d. diffuse

6. **Query** most nearly means
   a. bury
   b. wander
   c. ask
   d. praise

7. The thievery merited **severe** punishment.
   a. foul
   b. extreme
   c. cut
   d. shake

8. **Vexation** most nearly means
    a. boring
    b. pain
    c. revolving
    d. anger

9. The toddler, who had just learned to speak, seemed rather **loquacious**.
    a. verbose
    b. humorous
    c. silent
    d. cranky

10. **Grim** most nearly means
    a. frank
    b. dire
    c. corpse
    d. sharp

11. The bullies **disparaged** the younger boy, causing him to feel worthless.
    a. despaired
    b. belittled
    c. broke
    d. sparred

12. **Altitude** most nearly means
    a. behavior
    b. outlook
    c. haughtiness
    d. height

13. Though the valedictorian was very smart, he was too **egotistic** to have very many friends.
    a. conceited
    b. altruistic
    c. agrarian
    d. unconcerned

14. **Understand** most nearly means
    a. build
    b. fathom
    c. endow
    d. guilt

15. The car was jostled by the rocky **terrain**.
    a. firmament
    b. celestial
    c. arboreal
    d. ground

16. **Mitigate** most nearly means
   a. alleviate
   b. focus
   c. fiery
   d. conspire

17. The king's army was able to easily **encompass** his brother's army, and quickly crushed the rebellion.
   a. retreat
   b. brazen
   c. surround
   d. avoid

18. **Archetype** most nearly means
   a. ancestor
   b. literature
   c. policy
   d. model

19. After sleeping in his car for the past few months, the tiny hotel room seemed like a mansion by **contrast**.
   a. partisan
   b. variation
   c. revive
   d. improve

20. **Irate** most nearly means
   a. amiable
   b. innocent
   c. incensed
   d. tangible

21. Given his rude comments, her reaction seemed **plausible**.
   a. conceivable
   b. skeptical
   c. diurnal
   d. erudite

22. **Discourse** most nearly means
   a. praise
   b. disagreement
   c. speech
   d. path

23. The young boy seemed to be capable of nothing, other than his uncanny ability to **exasperate** his babysitter.
   a. annoy
   b. breathe
   c. disappoint
   d. stifle

24. **Bellicose** most nearly means
   a. loud
   b. angry
   c. pugnacious
   d. patriotic

25. Given the horrendous situation, the man's **equanimity** was appalling.
   a. justice
   b. hostility
   c. equine
   d. composure

26. **Preemptive** most nearly means
   a. beforehand
   b. prepare
   c. hidden
   d. initial

27. The rebels fought in order to **liberate** their brothers from the evil dictator.
   a. agitate
   b. instigate
   c. fracture
   d. release

28. **Mediocre** most nearly means
   a. excellent
   b. average
   c. intrusive
   d. inspiring

29. The jury was not impressed by the lazy defendant's **moot** argument.
   a. factual
   b. historical
   c. debatable
   d. exemplify

30. **Bestow** most nearly means
   a. take
   b. bequeath
   c. energize
   d. study

31. The starving man was disheartened when he reached the summit of the hill and realized that only a **barren** wasteland awaited him.
   a. fruitful
   b. infertile
   c. sumptuous
   d. lavish

32. **Refurbish** most nearly means
    a. renovate
    b. enlighten
    c. craven
    d. burnish

33. When she saw the crayon drawings on the wall, the mother had no choice but to **chastise** her sons.
    a. honor
    b. locate
    c. choose
    d. rebuke

34. **Sustain** most nearly means
    a. strength
    b. invigorate
    c. embolden
    d. expedite

35. When the boy saw how sincere the girl's apology was, he decided to **acquit** her of her faults.
    a. forgive
    b. stall
    c. acquire
    d. quit

Answer Explanations

1. A: Instigate refers to initiating or bringing about an action. While instigating something can be overwhelming, **overwhelm** describes the action, not what the action is. None of these choices match this except for **begin**, which is synonymous with instigate.

2. C: From the Latin root *fortis*, **fortify** literally means to make strong. **Foster** has nothing to do with the term at all. **Undermine** is to weaken, so this is the opposite. While a **barricade** can be used to fortify something, it is a means of protection, not the action of protection, unless used as a verb.

3. D: An **ultimatum** is a final demand or statement that carries a consequence if not met. The closest term presented is **threat** because, in context, an ultimatum carries some kind of punitive action or penalty if the terms are not met. Generally, it is usually used in a threatening or forceful context.

4. A: This was a tricky question because some choices connect **wave** to actions involved with water. **Sink** and **tide** are actions associated with the water, but are distinct from an ocean wave, which means wave is not used in context of an ocean wave, but as a verb to wave. To wave is to move around. **Stagnant** is to be still. **Flourish** describes a waving motion, making it the correct match.

5. C: It's easy to confuse void with **avoid**, which means keep away from. **Void** describes nothingness or emptiness. This means that **around** *(A)* is not the correct answer. Therefore, the best choice is **empty**, because this is the meaning of void.

6. C: Query comes from the Latin *querere*, which means to ask, so **query** means to ask a question. This Latin root also appears in the word **question**, which gives you a clue that the act of seeking answers is involved. **Ask** has the same meaning as query.

7. B: Severe reflects high intensity or very great. **Cut** and **shake** are verbs with actions that do not reflect severe. While something can be severely foul, **foul** is a broad description of something bad, but not necessarily of a high level of intensity. **Extreme** *(B)* describes something of the highest or most serious nature.

8. D: Vexation describes a state of irritation or annoyance. Think of the term vexed, meaning annoyed. The closest choice is **anger**, because vexation reflects annoyance, which, in most contexts, means a person is slightly angered or feels mild anger.

9. A: Loquacious reflects the tendency to talk a lot. While someone can be humorous and loquacious, **humorous** describes the kind of talk, not the fact that someone talks a lot. **Cranky** has nothing to do with the word, and **silent** is a clear opposite. This leaves **verbose**, meaning using an abundance of words; loquacious is a synonym.

10. B: Grim describes something of a dark or foreboding nature. **Frank** reflects sincerity, so does not relate. Neither does **corpse**, which is a body. **Sharp** describes a witty response or edged surface. This leaves **dire**, which is synonymous with grim. Both words reflect dark or unfruitful circumstances.

11. B: To **disparage** someone is to put them down. Although this may involve breaking a spirit and causing despair, **despair** is the result of disparage. To **spar** or fight may also be a result of disparage. **Belittle** means to bring someone down with words and is synonymous with disparage.

12. D: The other choices refer to **attitude**, a word spelled and pronounced similarly to altitude. Someone's attitude can reflect haughtiness, influence their outlook, and reflect their behavior. None of these terms describe **altitude**, which is the measure of height. **Height** is directly related to altitude.

13. A: Looking at the root word **ego**, egotistic must have something to do with the self—in this case, excessive self-interest. Such a person tends to be the opposite of **altruistic**, which means selfless. **Unconcerned** is also inappropriate, as egotistic people are concerned for themselves. **Agrarian** is an unrelated word concerning fields or farm lifestyle. **Conceited** is synonymous to egotistic.

14. B: Build and **guilt** can be ruled out because they are not related to **understand**. **Endow** is more difficult, because one can endow, or give, someone knowledge to understand, but endow involves the act of giving. **Fathom** is synonymous with understand; both terms reflect being able to grasp information.

15. D: Terrain, from the Latin *terra,* refers to the earth or physical landscape. **Celestial** and **firmament** both describe the sky. **Arboreal** describes things relating to trees. This leaves **ground**, another word for Earth or land, the same as terrain.

16. A: Mitigate refers to easing tension or making less severe. **Focus, fiery**, and **conspire** do not relate. **Alleviate** is synonymous, meaning to make less severe. Note the ending -*ate*, which also indicates function. Both terms reflect the function of easing difficult circumstances.

17. C: Encompass means to hold within or surround. **Avoid** and **retreat** are opposites. **Brazen** has no relation. **Surround**, meaning to **encircle**, is synonymous with **encompass**.

18. D: An **archetype** is an example of a person or thing, a recurrent symbol. An **ancestor** may be a good model of ideal behavior, but the term refers to someone who is related. **Archetypes** appear in literature, but these are different terms. **Policy** is unrelated. A **model** is a standard or representative, which is synonymous with archetype.

19. B: Contrast means to go against or to have a different perspective. **Revive** and **improve** are unrelated. **Partisan** can mean favoring one side, but variation is the best choice. **Variation** indicates clear difference, something that is not uniform and, therefore, contrasting.

20. C: Irate comes from the Latin *ira-*, which gives it the meaning of angry or irritable. **Amiable** means happy and friendly. **Innocent** and **tangible** have different and unrelated meanings. **Incensed** comes from the Latin *incedere*, which means to burn. Often this burning is a metaphor for extreme anger, the meaning of irate. Thus, incensed and irate are synonyms.

21. A: Plausible means likely to be possible or accepted, the opposite of **skeptical**. **Diurnal** relates to daytime, so is unrelated. **Erudite** means clever or intelligent, but not necessarily possible or correct. **Conceivable** means that something is able to be thought of or able to be done. Conceivable is the best match for plausible.

22. C: Discourse refers to spoken and written communication, or debate. **Path** doesn't have anything to do with discourse, unless used figuratively. While discourse may consist of praise or disagreement, all discourse—written or verbal—is a form of speech. Therefore, **speech** is the best term that encompasses the same meaning as discourse.

23. A: The Latin root, *asper*, means rough. **Exasperate**, then, means to make relations with someone rough or to rub them the wrong way. **Disappointment** can be the result an exasperating situation, but

these are results of the term, not the same as exasperation itself. **Annoy** is a synonym, meaning to irritate and make someone slightly angry.

24. C: The Latin root *bell*, which comes from *bellum*, refers to war. Someone who is **pugnacious** is ready for a fight. One may be **angry**, **patriotic**, or **loud**, but none of these terms directly relate to warlike behavior like pugnacious.

25. D: Someone who displays **equanimity**, like the *equ-* prefix suggests, is level-headed and even-tempered. While equine and equanimity appear to share the *equ-* prefix, **equine** refers to horses. **Justice** and **hostility** don't relate at all. This leaves **composure**, which also describes one's ability to keep a calm and level-headed state.

26. A: Preemptive refers to an action taken before an anticipated result can occur, often as a preventive measure. **Prepare** is similar—it's defined as actions taken before an event—but it doesn't necessarily involve preventative measures as does preemptive. Preemptive measures can be hidden, but that describes the act. **Initial**—meaning first—is close because it can be the first action in a series, but again, it doesn't refer to an action that is preventative. **Beforehand** is an action done in advance, before something occurs.

27. D: Agitate means to annoy, which is not the same as the meaning in this sentence. **Instigate** can mean to start, which is not synonymous with **liberate**. **Fracture** is to crack or break something, which can metaphorically be attributed to liberation (breaking of chains), but is not directly related to the word liberate. From the Latin root *liber*, meaning free, liberate means to free or release. **Release** is synonymous.

28. B: Mediocre is from the Latin *medius*, meaning middle. It refers to something of only decent quality, not exceptional. This rules out **excellent** and **inspiring**, as both communicate ideas of surpassing quality. **Intrusive** is unrelated. **Average** means usual or nothing out of the ordinary, like mediocre.

29. C: Moot means uncertain or in dispute. We can eliminate **factual, historical,** and **exemplify** because these are common terms with no ties to moot. **Debatable** is defined as having uncertain circumstances, leading to discussion or dispute.

30. B: Bestow means to give or present. **Take** can be eliminated because to take is the opposite action of to give. **Study** and **energize** mean different things entirely. **Bequeath** means to give or leave to another person.

31. B: Barren means deserted, void, lifeless, or having little. A desert is barren because it produces little vegetation. **Fruitful**, **sumptuous** and **lavish** express richness and abundance, which contradict barren. **Infertile** means unable to produce life, which mirrors barren.

32. A: Refurbish is to restore, set things up again, or make repairs. **Enlighten, craven,** and **burnish** are unrelated to these ideas. Note instead the *re-* prefix in **renovate**, which means again. This gives refurbish and renovate a meaning of restoring again, or returning to a better state. In other words, renewal.

33. D: Chastise means to reprimand severely. **Honor, choose,** and **locate** are unrelated. **Rebuke** is defined as harshly disapproving someone. Both rebuke and chastise are verbs, making rebuke a match.

34. B: Sustain is to revitalize, or to give strength. **Strength** would be a good choice, but strength alone does not describe the re-strengthening that sustain embodies. The best choice is **invigorate**, which is a verb like sustain, meaning to strengthen. This makes invigorate the better choice. **Expedite** is unrelated. While **embolden** means to give someone courage, which is a form of strength, invigorate and sustain speak more toward physical circumstances.

35. A: Acquit means to free of blame or charge. While acquire and acquit appear similar, they are unrelated. **Stall** and **quit** are also unrelated. **Forgive** literally means to pardon of sins or offenses. While not appearing related by their spellings, their meanings are nearly a perfect match, and they are both verbs.

Paragraph Comprehension

The **Paragraph Comprehension** subtest review section will cover:

- Comprehension Skills
- Purposes for Writing
- Writing Devices
- Types of Passages
- History and Culture in Relationship to Literature
- Responding to Literature
- Literary Genres
- Opinions, Facts, and Fallacies
- Organization of the Text
- Drawing Conclusions

Following the Paragraph Comprehension review material, a practice test along with a detailed answer key has been provided to help review the material covered throughout this section.

Comprehension Skills

Reading comprehension is the ability to read, process, and understand information. Reading comprehension skills such as **decoding** (the ability to use letter sound(s) to form and understand a word(s)), **fluency** (the pace at which one reads), **vocabulary knowledge** (the understanding of words one reads), and **background knowledge** (the use of familiarinformation to decipher unfamiliar texts) are basic building blocks effective readers use. Not only should effective readers utilize the basic comprehension building blocks, they should also use a variety of strategies to become better readers.

Good readers practice these comprehension skills before, during, and after reading. Before readers go through a text or passage, they should ask the following questions: "What do I already know about this subject?" and "What will the book be about?" While reading the text or passage, readers should visualize the text, use new information to make predictions, and continue to self-question. Furthermore, readers should also make connections to a personal experience or another event that occurred within other texts. After reading a text or passage, readers should be able to summarize and process what the main idea of the text or passage was about. Good readers are able to retell the story in their own words.

The Comprehension Skills section covers six types of comprehension strategies, the topic versus the main idea, supporting details, and the theme. The chart on the following page shows the six common types of comprehension strategies: question, connect, infer, visualize, what's important, and synthesize.

Six Types of Comprehension Strategies

1.

Question

Monitor reading by asking questions before, during, and after reading a text. "What if?," "I don't understand why," or "Maybe when" are all examples of how to utilize questioning while reading.

2.

Connect

Use knowledge to help with understanding text. Connecting with other texts or experiences is a way to help with text comprehension." This reminds me of… because" and "When I heard…, it reminds me of" are both examples of how to connect with texts.

3.

Infer

When authors do not give a clear answer, it is often necessary to infer what happens in the text. Inferring helps with making predictions, drawing conclusions, and reflecting on text. "I think," "Maybe," and "Perhaps" are all examples of ways to infer while reading.

4.

Visualize

Create pictures in the mind about the text. "I see," "It must have smelled like," and "I can imagine" are all examples of how to visualize while reading.

5.

What's important

Determine the author's main idea. "The main idea is," "This section is primarily about," and "It is important to remember" are all examples of how to determine what is important while reading.

6.

Synthesize

Combine current knowledge with new information to help with text understanding. "After reviewing," "At first I thought," and "Now I think" are all examples of how to synthesize information while reading.

Topic Versus Main Idea

It is very important to know the difference between the topic and the main idea of the text. Even though these two are similar because they both present the central point of a text, they have distinctive differences. A **topic** is the subject of the text; it can usually be described in a one- to two-word phrase and appears in the simplest form. On the other hand, the **main idea** is more detailed and provides the author's central point of the text. It can be expressed through a complete sentence and is often found in the beginning, middle, or end of a paragraph. In most nonfiction books, the first sentence of the passage usually (but not always) states the main idea. Review the passage below to explore the topic versus the main idea.

> Cheetahs are one of the fastest mammals on the land, reaching up to 70 miles an hour over short distances. Even though cheetahs can run as fast as 70 miles an hour, they usually only have to run half that speed to catch up with their choice of prey. Cheetahs cannot maintain a fast pace over long periods of time because their bodies will overheat. After a chase, cheetahs need to rest for approximately 30 minutes prior to eating or returning to any other activity.

In the example above, the topic of the passage is "Cheetahs" simply because that is the subject of the text. The main idea of the text is "Cheetahs are one of the fastest mammals on the land but can only maintain a fast pace for shorter distances." While it covers the topic, it is more detailed and refers to the text in its entirety. The text continues to provide additional details called supporting details, which will be discussed in the next section.

Supporting Details

Supporting details help readers better develop and understand the main idea. Supporting details answer questions like *who, what, where, when, why,* and *how*. Different types of supporting details include examples, facts and statistics, anecdotes, and sensory details.

Persuasive and informative texts often use supporting details. In persuasive texts, authors attempt to make readers agree with their points of view, and supporting details are often used as "selling points." If authors make a statement, they need to support the statement with evidence in order to adequately persuade readers. Informative texts use supporting details such as examples and facts to inform readers. Review the previous "Cheetahs" passage to find examples of supporting details.

> Cheetahs are one of the fastest mammals on the land, reaching up to 70 miles an hour over short distances. Even though cheetahs can run as fast as 70 miles an hour, they usually only have to run half that speed to catch up with their choice of prey. Cheetahs cannot maintain a fast pace over long periods of time because their bodies will overheat. After a chase, cheetahs need to rest for approximately 30 minutes prior to eating or returning to any other activity.

In the example, supporting details include:

- Cheetahs reach up to 70 miles per hour over short distances.
- They usually only have to run half that speed to catch up with their prey.
- Cheetahs will overheat if they exert a high speed over longer distances.
- Cheetahs need to rest for 30 minutes after a chase.

Look at the diagram below (applying the cheetah example) to help determine the hierarchy of topic, main idea, and supporting details.

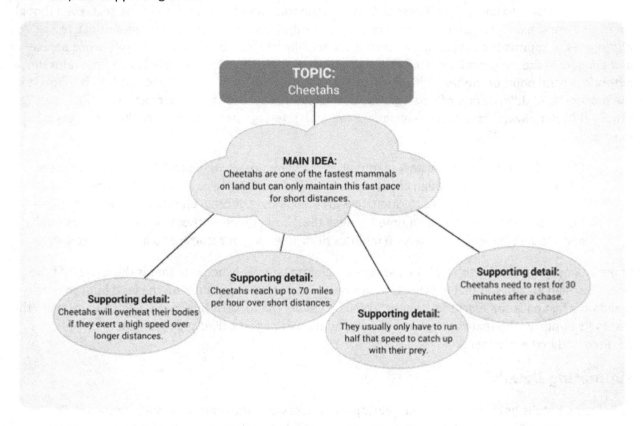

Theme

The **theme** of a text is the central message of the story. The theme can be about a moral or lesson that the author wants to share with the audience. Although authors do not directly state the theme of a story, it is the "big picture" that they intend readers to walk away with. For example, the fairy tale *The Boy Who Cried Wolf* features the tale of a little boy who continued to lie about seeing a wolf. When the little boy actually saw a wolf, no one believed him because of all of the previous lies. The author of this fairy tale does not directly tell readers, "Don't lie because people will question the credibility of the story." The author simply portrays the story of the little boy and presents the moral through the tale.

The theme of a text can center around varying subjects such as courage, friendship, love, bravery, facing challenges, or adversity. It often leaves readers with more questions than answers. Authors tend to insinuate certain themes in texts; however, readers are left to interpret the true meaning of the story.

Purposes for Writing

Authors want to capture the interest of the reader. An effective reader is attentive to an author's position. Authors write with intent, whether implicit or explicit. An author may hold a bias or use emotional language, which, in turn, creates a very clear position. Determining an author's purpose is usually easier than figuring out his or her position. An author's purpose for a text may be to persuade, inform, entertain, or be descriptive. Most narratives are written with the intent to entertain the reader, although some may also be informative or persuasive. When an author tries to persuade a reader, the reader must be cautious of the intent or argument. Therefore, authors keep the persuasion lighthearted

and friendly to maintain the entertainment value in narrative texts even though they are still trying to convince the reader of something.

An author's purpose will influence their writing style. As mentioned previously, the purpose can inform, entertain, or persuade a reader. If an author writes an **informative text**, his or her purpose is to educate the reader about a certain topic. Informative texts are usually nonfiction, and the author rarely states his or her opinion. The purpose of an informative text is also indicated by the outline of the text itself. In some cases, an informative text may have headings, subtitles, and bold key words. The purpose for this type of text is to educate the reader.

Entertaining texts, whether fiction or nonfiction, are meant to captivate readers' attention. Entertaining texts are usually stories that describe real or fictional people, places, or things. These narratives often use expressive language, emotions, imagery, and figurative language to captivate the readers. If readers do not want to put the entertaining text down, the author has fulfilled his or her purpose for this type of text.

Descriptive texts use adjectives and adverbs to describe people, places, or things to provide a clear image to the reader throughout the story. If an author fails to provide detailed descriptions, readers may find texts boring or confusing. Descriptive texts are almost always informative but can also be persuasive or entertaining, depending the author's purpose.

Writing Devices

Authors use a variety of writing devices throughout texts. Below is a list of some of the stylistic writing devices authors use in their writing:

- Comparison and Contrast
- Cause and Effect
- Analogy
- Point of View
- Transitional Words and Phrases

Comparison and Contrast

One writing device authors use is comparison and contrast. When authors take two objects and show how they are alike or similar, a **comparison** is being made. When authors take the same two objects and show how they differ, they are **contrasting** them. Comparison and contrast essays are most commonly

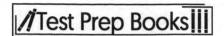

written in nonfiction form. A review of the Venn diagram demonstrating common words or phrases used when comparing or contrasting objects appears below.

Compare and Contrast Venn Diagram

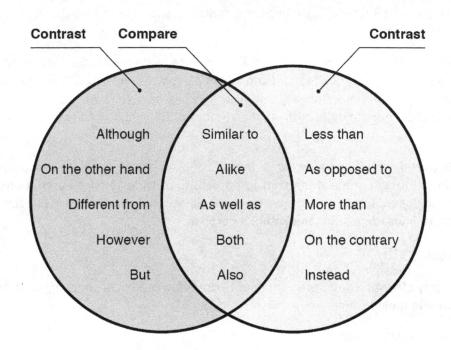

Cause and Effect

Cause and effect is one of the most common writing devices. A **cause** is why something happens, whereas an **effect** is the result that occurs because of the cause. Oftentimes, authors use key words to show cause and effect, such as *because, so, therefore, without, now, then,* and *since*. For example,

> *Because* of the sun shower, a rainbow appeared.

In this sentence, due to the sun shower (the cause), a rainbow appeared (the effect).

Analogy

An **analogy** is a comparison between two things that are quite different from one another. Authors commonly use analogies to add meaning and make ideas relatable in texts. Metaphors and similes are specific types of analogies. Metaphors compare two things that are not similar and directly connect them. Similes also compare two unlike items but connect them using the words *like* or *as*. For example,

> In the library, Alice was asked to be as quiet as a mouse.

Clearly, Alice and a mouse are very different. However, when Alice is asked to be as quiet as a mouse, readers understand that mice are small and therefore have small and soft voices—appropriate voice noise level for the library.

Point of View

Point of view is the perspective in which authors tell stories. Authors can tell stories in either the first or third person. When authors write in the first person, they are a character within a story telling about their own experiences. The pronouns *I* and *we* are used when writing in the first person. If an author writes in the third person, the narrator (the person telling the story) is telling the story from an outside perspective and is completely detached from the story. The author is not a character in the story, but rather tells about the characters' actions and dialogues. Pronouns such as *he, she, it,* and *they* are used in texts written in the third person.

Transitional Words and Phrases

There are approximately 200 transitional words and phrases that are commonly used in the English language. Below are lists of common transition words and phrases used throughout transitions.

Time
- after
- before
- during
- in the middle

Example about to be Given
- for example
- in fact
- for instance

Compare
- likewise
- also

Contrast
- however
- yet
- but

Addition
- and
- also
- furthermore
- moreover

Logical Relationships
- if
- then

- therefore
- as a result
- since

Steps in a Process
- first
- second
- last

Transitional words and phrases are important writing devices because they connect sentences and paragraphs. Transitional words and phrases present logical order to writing and provide more coherent meaning to readers.

Types of Passages

Authors write with different purposes in mind. They use a variety of writing passages to appeal to their chosen audience. There are four types of writing passages:

- Narrative
- Expository
- Technical
- Persuasive

Each one will be described individually.

Narrative
Narrative writing tells a story or a series of events. A narrative can either be fiction or nonfiction, although in order to still be categorized as a narrative, certain elements need to be present.

A narrative must have:

- **Plot**: what happens in the story, or what is going to happen
- **Series of events**: beginning, middle, and end, but not necessarily in that order
- **Characters**: people, animals, or inanimate objects
- **Figurative language**: metaphors, similes, personification, etc.
- **Setting**: when and where the story takes place

Expository
Expository passages are informative texts usually written as memoirs or autobiographies. These nonfiction passages often use transitional words and phrases like *first, next,* and *therefore* to provide readers with a clear sense of direction of where they are within the text. Because expository passages are meant to educate readers, they often do not use flamboyant language, unless the subject area requires it.

Technical
Technical passages are written to describe how to do or make something. Technical passages are often manuals or guides written in a very organized and logical manner. The texts usually have outlines with subtitles and very little jargon. The vocabulary used in technical passages is very straightforward so as

not to confuse readers. Technical texts often explore cause and effect relationships and can also include authors' purposes.

Persuasive

Persuasive passages are written with the intent to change readers' opinions to agree with the author's viewpoint on the subject. Authors generally stick to one main point and present smaller arguments that concur with the initial central claim. The author's intent or purpose is to persuade readers by presenting them with evidence or clues (such as facts, statistics, and observations) to make them stray away from their own thoughts on the matter and agree with the author's ideas. By using evidence and clues to support their ideas, authors may or may not be strategically placing strong thoughts and emotions in their readers' minds, causing them to lean in one direction more than the other.

Readers' opinions are formed from these strong thoughts and emotions. Authors need to be wary of misleading their readers because it is not ethical, nor is it appropriate. In the same sense, readers also need to be skeptical of an author's intent if the persuasion comes from an emotional standpoint.

For example, during an election year, candidates often post slanderous advertisements about one another. These malicious ads do not state what the candidate will do for his or her country or district. They are meant to show why readers or viewers should not elect the other candidate. These ads are trying to persuade viewers' opinions of opposing candidates. On the other hand, there are other ads that show the positive impacts that candidates have made. These types of advertisements also persuade voters into liking those candidates because of the positive claims portrayed. Are these advertisements right or wrong? Are they ethical? It is up to the viewer or reader to decide.

History and Culture in Relationship to Literature

Historical events and different cultures have strong influences on literature. Texts written during the Great Depression differ greatly from texts written during the Vietnam War because of the historical context. Time periods and events inspire a variety of topics and provide authors with creative liberty to express their opinions directly or inadvertently. Although modern readers may recognize an author's prejudices from the past, it is important to place the writing in relation to the time period from which it came.

Despite varying historical events, authors generally have universal themes. Themes such as independence and self-control, individual versus society, personal growth and strife, and heroes and heroines are present in literature of all cultures. Authors from various places have also used similar genres for centuries, such as epic poetry, drama, or satire. For example, England and Greece have both expressed controversial ideas or theories from a religious standpoint using satire. In some cultures, authors write plays or theatrical pieces to express opposing viewpoints, whereas in other cultures, authors create poems to create strong allusions and imagery for readers. Overall, authors throughout different time periods and cultures use universal themes but express their writing through various writing forms.

Responding to Literature

When writing, authors' intentions are to captivate their readers. Good literature draws readers into books and keeps them actively involved throughout the text. Part of being an active and engaged reader is making predictions, making inferences, finding the sequence of events, summarizing, and drawing conclusions.

A **prediction** is a forecast of what will happen next. Good readers constantly make predictions based on what they already know or read. As readers are going through texts, information is shared so that one can then assume what may happen next; granted, readers' predictions may not always be correct. For example:

> Kendall heard a scratching sound at the closet door. As she got closer, a steady purring noise got louder and louder.

Readers can predict that a cat may be behind the closet door. However, it is the readers' job to determine what the character(s) may do next. In this example, will Kendall open the door, and if so, what will the cat's response be? Authors may give the readers more clues leading up to the actual reveal, but the readers' job is to start making predictions about what may come next.

Like predictions, readers can also make inferences. An **inference** is when readers can reach a conclusion based upon the other information given. Authors do not directly state the outcome, nor do they imply what actually happened. Inferences allow readers to analyze information to come to a conclusion. In the sentence, *Molly collected seashells and then built a sand castle*, readers can infer that Molly is at the beach. Readers can use their prior knowledge, context clues, and facts to help them build inferences.

Sometimes readers' inferences may be true but incorrect at the same time. Readers must be conscious of understanding the context in which the clues are stated. Sometimes, authors may mean something figuratively and not literally, or vice versa.

Along with making predictions and inferences, active and engaged readers also identify the **sequence of events** in texts. Oftentimes, authors use words like *first, then, next,* and *last* to identify the chronological order of stories. However, authors do not always use transitional words if the sequence of events is already implied. For example, in the sentence: *Tom plugged in his iPod, and it started to charge*, readers can infer that Tom's iPod would not charge if it had not been plugged in first. The fact that the natural order of events is already implied is the reason the author does not use any transitional words to guide readers.

Finally, after readers have made predictions and inferences and identified the sequence of events, they must **summarize** and/or **draw conclusions** about what they read. If readers cannot retell what they read, then they do not fully comprehend the text. Readers can identify authors' conclusions or summaries by some key transitional words such as *finally, overall,* or *in conclusion*. When summarizing passages, readers should be careful not to use definite phrases such as *always* or *never* because those words do not leave any room for exceptions. Also, when drawing conclusions about passages, readers should only use the information provided in the passages. This is not a time to use any related information; instead, they should stick to the facts.

Literary Genres

Literary genres allow literature to be classified and categorized. The basic literary genres are poetry, fiction, nonfiction, and drama. Within each literary genre category, numerous subgroups are found. At times, genres overlap, which can be confusing to readers. However, identifying texts' genres can help readers develop a better understanding to guide their responses to texts.

Poetry
Poetry is often thought to be one of the oldest forms of literature. Poetry uses patterns of language like rhyming, counting syllables or lines, or free-flowing expressive words, structured lines, or stanzas.

Figurative language (such as metaphors and imagery) is also used in poetry. Poems are intended to appeal to readers' emotions and make readers think and feel through the words. Various types of poems include, but are not limited to:

- **Rhyming/metrical**: A poem that rhymes to relay messages in a metrical pattern.

- **Haiku**: Traditional Japanese poetry that is designed as a three-line poem. The first line has five syllables, the second line has seven syllables, and the third and final line has five syllables.

- **Sonnet**: A fourteen-line poem where each line has approximately ten syllables of iambic pentameter (alternating in stressed and unstressed patterns).

- **Free verse**: A poem without limitations; there may or may not rhyming throughout the poem.

Fiction

Fictional texts, by definition, are not real. Plots, settings, and characters are invented by authors. Some authors use real events or people but twist the characters and plots of stories, making it **prose fiction**. At times the line between fiction and poetry blurs with pieces of **non-prose fiction** such as songs, ballads, epics, and narrative poems. There are several subgroups of prose fiction that include:

- **Short story**: A piece of fiction meant to be read in one sitting. A short story typically has just a few characters and a basic plot, which mainly describes one major event. Short stories originated from storytelling traditions in the seventeenth century.

- **Novel**: A lengthy fictitious narrative that has a complex plot and characters, and it may have a sequential order of events and action scenes.

- **Novella**: A short fictitious novel or long short story. Novellas originated from the German culture.

- **Myth**: A traditional fictitious story, usually about ancient history, social marvel, or supernatural events. Myths stem back to Greek mythos.

- **Fable**: A fictitious short story, usually about animals, that is intended to teach a moral or lesson. Animals in fables usually have human-like characteristics, such as the ability to speak.

There are many components that influence prose fiction including:

- **Speech** and **dialogue**: Characters in prose fiction may carry the dialogue through a narrator or themselves. The dialogue itself can be based off reality or fantasy, depending on the author's purpose.

- **Thoughts** and **mental processes:** Sometimes authors use an internal conversation to help develop the plot or characters.

- **Dramatic involvement**: At times, narrators encourage readers to be involved in the storyline. On the other hand, some narrators urge readers to distance themselves from the events of the story.

- **Action**: The actual events and information that keep the story's plots and character development moving forward.

- **Duration:** The period of time, either long or short over which the events in the story take place. Duration varies between the relationship of described and narrative time.

- **Setting** and **description:** The location or timeframe (setting) important to the plot, character development, and how events are being described.

- **Theme:** The overall topic or event that is addressed through the story.

- **Symbolism:** The use of imagery and other literary devices to reveal the author's true intent.

Nonfictional pieces of prose originate from facts but contain fictional elements. They may be used to teach, persuade, or share experiences of reality. Essays and biographies fall under this category. As another example, **historical fiction** uses a factual period of time, person, or place, but may embellish or invent details for the sake of the narrative.

Whether readers are reading fictional or nonfictional prose, it is important that they remember to analyze the language being used throughout the text. The language of fiction is not only a way to connect with the plot or characters, but it also gives readers a variety of ways to relate to familiar emotions, events, or objects. Despite authors' rigorous language, readers reap the rewards at the end.

As mentioned, dialogue can be expressed via narrators or characters. The **narrator** can be thought of as a storyteller. Narrators deliver dialogue and insight about main ideas, themes, and characters in stories. To help readers better understand the purpose and role of narrators in fictitious stories, it is important that they ask themselves the following questions:

- Who is the narrator, and what person perspective (first or third) are they speaking in?

- Is the narrator a character in the story, or is the narrator simply explaining events in the text? Sometimes narrators outwardly approve or disapprove of events, plots, or character development in stories.

- What is the narrator's tone, and who is the narrator's audience? Is the narrator speaking in an informal or a formal manner? Does the author's vocabulary give any clues about the narrator?

Aside from narrators, **characters** also create dialogue to develop the plot, mood, or message of the story. Characters play an important role in the plot development simply because they are the ones who move the storyline along. Authors usually provide detailed descriptions of characters, but not always. They often provide details such as physical qualities, psychological state, or a character's motivation to help readers determine the role the characters play in the given story.

Characters are said to be either flat or round. **Flat characters** are not key members of the story. They are minor figures in the storyline and do not adjust the overall plot or events. On the other hand, **round characters** are important to the plot and tend to change as the story unfolds. Characters can maintain internal dialogue, providing readers with a deeper understanding of characters' relationships.

Nonfiction

Nonfiction refers to texts that tell about real events or inform, teach, and explain information to readers. Unlike fiction texts, nonfiction texts are about real people, places, and things. Nonfiction texts may include charts, diagrams, photographs, a table of contents, an index, labels, glossaries, and labels.

Authors have intended audiences for nonfiction pieces; therefore, there are many types or subgroups of nonfiction texts, such as:

- **Autobiography**: A true story where authors write about themselves and their own life.

- **Biography**: A true story where authors write about someone else's life (not their own).

- **News**: Information, usually important or recent events, worthy of being broadcasted or published.

- **Journal**: A publication, usually a magazine or newspaper, that provides detailed information about specific subjects or events.

- **Essay**: A short piece of writing on a particular subject. (Essays can also be fiction.)

- **Textbook**: A book used to inform and educate readers on a particular subject.

Authors can also provide additional information in nonfiction texts by using style, tone, perspective, and purpose. Authors' styles show a little bit of their personalities. Whether it is the language authors choose to use or the organization of text, their styles vary.

An author's **tone** is one way in which the author presents their attitudes to the intended audience. Whether an author's words are friendly, serious, arrogant, or good-humored, readers can "hear" tones of authors through words. The tones can influence how texts are interpreted.

Viewpoint, or **perspective**, points out whether or not authors hold biases about subjects. Authors may directly or indirectly present opinions on particular subjects by using strong, emotional language or by simply choosing to ignore important pieces of information to strengthen the case for their "side."

As always, authors have a purpose when they write. They write to inform, educate, influence, entertain, and possibly persuade their readers.

Drama

Drama is another important genre. **Drama** is written with the intention of being performed in a play/theater or on radio or television. Although dramas can also be read, they can be performed as plays (live or recorded audiences), soliloquies (when someone speaks about his or her thought(s) aloud and disregards who can hear), asides (parts of actors' lines not heard by others on stage and only intended to be heard by the audience), or different modes of dialogue. The variety of dramatic devices provides audiences or readers with lots of hints to help them identify the major themes or plots in stories.

Drama is meant to be highly entertaining, emotional, and exciting for the intended audiences. It can inspire and change people who watch or read it. Plots often revolve around real-life situations, social problems, conflict/resolution, illnesses, or addictions, and events occur in a predictable pattern. The diagram of "Freytag's Pyramid" demonstrates the sequence of events that the plots of most dramas and fictional pieces follow. In Freytag's Pyramid, **exposition** refers to the beginning or early stages of the drama. It sets the theme and setting and introduces major characters. The **rising action** is when conflict starts and tensions arise. The **climax** is the moment that is also considered the crisis moment; it is the time where there is the greatest amount of tension. The **falling action** is when the final catastrophe occurs. Finally, the **resolution** is where the piece finds closure.

The action in dramatic pieces is crucial. Plays that have little dialogue and a lot of action are common. Stage directions are important to help understand where characters are at all times, to determine who is speaking versus who is listening, and to follow patterns throughout scenes.

As mentioned previously, dramas are often written with the intention that they will be performed but they can also be read through texts. There are some benefits to watching a play in a theater versus reading dramas in texts. If dramas are not performed in plays, authors have to provide lots of details so that readers can imagine the setting and characters. Readers should be able to visualize how characters interact and how the plot unfolds. However, their interpretation of people or places may be different from the author's intent. Whereas, if an audience watched a play demonstrating the same exact storyline that they read, there would not be any confusion because they can visually see the setting, characters, and plot unfold. In this sense, there is no room for interpretation.

For example, with Shakespeare's drama *Romeo and Juliet*, based on language alone, readers may get confused, given that the language used is from a completely different time period. On the other hand, if an audience watches Shakespeare's play *Romeo and Juliet,* they are able to see the actors' emotions and the setting instead of trying to interpret the language.

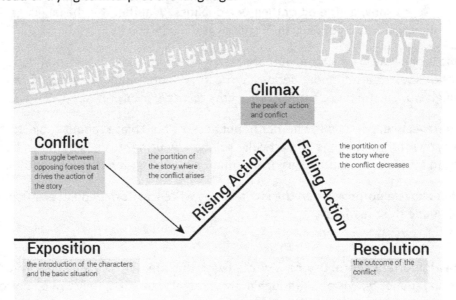

Whether they are read or performed/watched, dramas have three critical elements: dialect, speech, and dialogue. Each of these key elements affects one another. Characters in dramas have a unique way of speaking, whether it is through the tone, language, or emphasis used. The interpretation of characters' lines depends on some of these critical elements.

Dialect is the variety of language used by a particular group, especially one spoken in a specific region or part of the county. There can be many dialects among the same language.

For example, in France, the spoken language is French. Parts of Haiti also speak French. Just because the two countries speak the same language does not mean they use the same dialect. Therefore, individuals from France and Haiti may or may not understand each other, or they might misinterpret one another in conversation.

Combinations of dialects are one way of forming speech. Neighboring countries' cultures and dialects naturally influence and cross over borders. **Dialect geography** is the study of speech differences from

one geographic area to another. Atlases visually demonstrate where dialects and speech cross over or run independently from one another. A dialect continuum demonstrates how dialects move across regions.

The tone of **dramatic dialogue** drastically changes how characters and audiences interpret lines or roles. The emphasis of certain words, phrases, or actions determines the meaning behind the dramatic dialogue. Dramatic speeches can use various types of tone, especially when expressing feelings or emotions, to manipulate the direction of the speech itself. Readers better understand characters once they decipher the tone(s) being used throughout the dialogue.

Throughout all literary genres, art and music can enhance ways in which audiences, authors, and performers interact with different texts or pieces. Illustrations, dance, fashion, architecture, and/or musical scores can add meaning to imagery and motion that words simply cannot do. When art and/or music are added to literary genres, it is important to keep everything relevant to the coordinating time period. For example, a jazz piece with saxophones and trombones should not be played while a Shakespearian piece is being read or performed as this would be an anachronism. By adding art or music to literary pieces, the overall experience is enhanced.

Opinions, Facts, and Fallacies

As mentioned previously, authors write with a purpose. They adjust their writing for an intended audience. It is the readers' responsibility to comprehend the writing style or purpose of the author. When readers understand a writer's purpose, they can then form their own thoughts about the text(s) regardless of whether their thoughts are the same as or different from the author's. The following section will examine different writing tactics that authors use, such as facts versus opinions, bias and stereotypes, appealing to the readers' emotions, and fallacies (including false analogies, circular reasoning, false dichotomy, and overgeneralization).

Facts Versus Opinions

Readers need to be aware of the writer's purpose to help discern facts and opinions within texts. A **fact** is a piece of information that is true. It can either prove or disprove claims or arguments presented in texts. Facts cannot be changed or altered. For example, the statement: *Abraham Lincoln was assassinated on April 15, 1865*, is a fact. The date and related events cannot be altered.

Authors not only present facts in their writing to support or disprove their claim(s), but they may also express their opinions. Authors may use facts to support their own opinions, especially in a persuasive text; however, that does not make their opinions facts. An **opinion** is a belief or view formed about something that is not necessarily based on the truth. Opinions often express authors' personal feelings about a subject and use words like *believe, think,* or *feel.* For example, the statement: *Abraham Lincoln was the best president who has ever lived*, expresses the writer's opinion. Not all writers or readers agree or disagree with the statement. Therefore, the statement can be altered or adjusted to express opposing or supporting beliefs, such as "Abraham Lincoln was the worst president who has ever lived" or "I also think Abraham Lincoln was a great president."

When authors include facts and opinions in their writing, readers may be less influenced by the text(s). Readers need to be conscious of the distinction between facts and opinions while going through texts. Not only should the intended audience be vigilant in following authors' thoughts versus valid information, readers need to check the source of the facts presented. Facts should have reliable sources derived from credible outlets like almanacs, encyclopedias, medical journals, and so on.

Bias and Stereotypes

Not only can authors state facts or opinions in their writing, they sometimes intentionally or unintentionally show bias or portray a stereotype. A **bias** is when someone demonstrates a prejudice in favor of or against something or someone in an unfair manner. When an author is biased in his or her writing, readers should be skeptical despite the fact that the author's bias may be correct. For example, two athletes competed for the same position. One athlete is related to the coach and is a mediocre athlete, while the other player excels and deserves the position. The coach chose the less talented player who is related to him for the position. This is a biased decision because it favors someone in an unfair way.

Similar to a bias, a **stereotype** shows favoritism or opposition but toward a specific group or place. Stereotypes create an oversimplified or overgeneralized idea about a certain group, person, or place. For example,

> Women are horrible drivers.

This statement basically labels *all* women as horrible drivers. While there may be some terrible female drivers, the stereotype implies that *all* women are bad drivers when, in fact, not *all* women are. While many readers are aware of several vile ethnic, religious, and cultural stereotypes, audiences should be cautious of authors' flawed assumptions because they can be less obvious than the despicable examples that are unfortunately pervasive in society.

Appealing to the Readers' Emotions

Authors write to captivate the attention of their readers. Oftentimes, authors will appeal to their readers' emotions to convince or persuade their audience, especially when trying to win weak arguments that lack factual evidence. Authors may tell sob stories or use bandwagon approaches in their writing to tap into the readers' emotions. For example, "Everyone is voting yes" or "He only has two months to live" are statements that can tug at the heartstrings of readers. Authors may use other tactics, such as name-calling or advertising, to lead their readers into believing something is true or false. These emotional pleas are clear signs that the authors do not have a favorable point and that they are trying to distract the readers from the fact that their argument is factually weak.

Fallacies

A **fallacy** is a mistaken belief or faulty reasoning, otherwise known as a **logical fallacy**. It is important for readers to recognize logical fallacies because they discredit the author's message. Readers should continuously self-question as they go through a text to identify logical fallacies. Readers cannot simply complacently take information at face value. There are six common types of logical fallacies:

- False analogy
- Circular reasoning
- False dichotomy
- Overgeneralization
- Slippery slope
- Hasty generalization

Each of the six logical fallacies are reviewed individually.

False Analogy

A **false analogy** is when the author assumes two objects or events are alike in all aspects despite the fact that they may be vastly different. Authors intend on making unfamiliar objects relatable to convince readers of something. For example, the letters *A* and *E* are both vowels; therefore, *A* = *E*. Readers cannot assume that because *A* and *E* are both vowels that they perform the same function in words or independently. If authors tell readers, *A* = *E*, then that is a false analogy. While this is a simple example, other false analogies may be less obvious.

Circular reasoning

Circular reasoning is when the reasoning is decided based upon the outcome or conclusion and then vice versa. Basically, those who use circular reasoning start out with the argument and then use false logic to try to prove it, and then, in turn, the reasoning supports the conclusion in one big circular pattern. For example, consider the two thoughts, "I don't have time to get organized" and "My disorganization is costing me time." Which is the argument? What is the conclusion? If there is not time to get organized, will more time be spent later trying to find whatever is needed? In turn, if so much time is spent looking for things, there is not time to get organized. The cycle keeps going in an endless series. One problem affects the other; therefore, there is a circular pattern of reasoning.

False dichotomy

A **false dichotomy**, also known as a false dilemma, is when the author tries to make readers believe that there are only two options to choose from when, in fact, there are more. The author creates a false sense of the situation because he or she wants the readers to believe that his or her claim is the most logical choice. If the author does not present the readers with options, then the author is purposefully limiting what readers may believe. In turn, the author hopes that readers will believe that his or her point of view is the most sensible choice. For example, in the statement: *you either love running, or you are lazy*, the fallacy lies in the options of loving to run or being lazy. Even though both statements do not necessarily have to be true, the author tries to make one option seem more appealing than the other.

Overgeneralization

An **overgeneralization** is a logical fallacy that occurs when authors write something so extreme that it cannot be proved or disproved. Words like *all, never, most,* and *few* are commonly used when an overgeneralization is being made. For example,

> All kids are crazy when they eat sugar; therefore, my son will not have a cupcake at the birthday party.

Not *all* kids are crazy when they eat sugar, but the extreme statement can influence the readers' points of view on the subject. Readers need to be wary of overgeneralizations in texts because authors may try to sneak them in to sway the readers' opinions.

Slippery slope

A **slippery slope** is when an author implies that something will inevitably happen as a result of another action. A slippery slope may or may not be true, even though the order of events or gradations may seem logical. For example, in the children's book *If You Give a Mouse a Cookie*, the author goes off on tangents such as "If you give a mouse a cookie, he will ask for some milk. When you give him the milk, he'll probably ask you for a straw." The mouse in the story follows a series of logical events as a result of a previous action. The slippery slope continues on and on throughout the story. Even though the mouse made logical decisions, it very well could have made a different choice, changing the direction of the story.

Hasty generalization

A **hasty generalization** is when the reader comes to a conclusion without reviewing or analyzing all the evidence. It is never a good idea to make a decision without all the information, which is why hasty generalizations are considered fallacies. For example, if two friends go to a hairdresser and give the hairdresser a positive recommendation, that does not necessarily mean that a new client will have the same experience. Two referrals is not quite enough information to form an educated and well-formed conclusion.

Overall, readers should carefully review and analyze authors' arguments to identify logical fallacies and come to sensible conclusions.

Organization of the Text

The **structure of the text** is how authors organize information in their writing. The organization of text depends on the authors' intentions and writing purposes for the text itself. Text structures can vary from paragraph to paragraph or from piece to piece, depending on the author. There are various types of patterns in which authors can organize texts, some of which include problem and solution, cause and effect, chronological order, and compare and contrast. These four text structures are described below.

Problem and Solution

One way authors can organize their text is by following a **problem and solution** pattern. This type of structure may present the problem first without offering an immediately clear solution. The problem and solution pattern may also offer the solution first and then hint at the problem throughout the text. Some texts offer multiple solutions to the same problem, which then leaves the decision as to the "best" solution to the minds of readers.

Even though the problem and solution text may seem easy to recognize, it is often confused with another organizational text pattern, cause and effect. One way of determining the difference between the two patterns is by searching for key words that indicate a problem and solution organizational text pattern is being followed, such as *propose, answer, prevention, issue, fix,* and *problematic.* Also, in problem and solution patterned texts, solutions are offered to all problems, even negative problems, unlike the cause and effect pattern.

Cause and Effect

Cause and effect is one of the more common ways that authors organize texts. In a cause and effect patterned text, the author explains what caused something to happen. For example, "It rained, so we got all wet." In this sentence, the cause is "the rain," and the effect is "we got all wet." Authors tend to use key words such as *because, as a result, due to, effected, caused, since, in order,* and *so* when writing cause and effect patterned texts.

Persuasive and expository writing models frequently use a cause and effect organizational pattern as well. In **persuasive texts**, authors try to convince readers to sway their opinions to align with the authors' thoughts. Authors may use cause and effect patterns or relationships to present supporting evidence to try to persuade the readers' judgment on a particular subject.

In **expository texts**, authors write to inform and educate readers about certain subjects. By using cause and effect patterns in expository writing, authors show relationships between events—basically, how one event may affect the other in chronological order.

Chronological Order

When using a **chronological order** organizational pattern, authors simply state information in the order in which it occurs. Nonfiction texts often include specific dates listing events in chronological order (such as timelines), whereas fiction texts may list events in order but not provide detailed dates (such as describing a daily routine: wake up, eat breakfast, get dressed, and so on). Narratives usually follow the chronological order pattern of beginning, middle, and end, with the occasional flashback in between.

Compare and Contrast

The **compare and contrast** organizational text pattern explores the differences and similarities of two or more objects. If authors describe how two or more objects are similar, they are comparing the items. If they describe how two or more objects are different, they are contrasting them. In order for texts to follow a compare and contrast organizational pattern, authors must include both similarities and differences within the text and hold each to the same guidelines. The following Venn diagram compares and contrasts oranges and apples.

Compare and Contrast Example

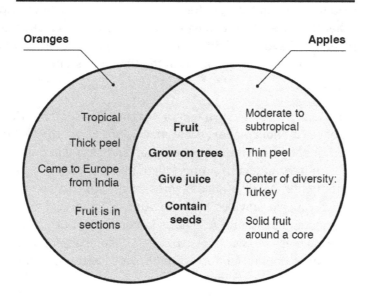

In this diagram, readers should notice how the author uses the same guidelines when comparing and contrasting oranges and apples. The origin, peel type, climate, outcomes, and classifications are all used for both fruits. The author does not use different categories for each fruit to compare or contrast the two fruits.

Even though authors may agree or disagree with each side of an argument, they should remain impartial when presenting the facts in a compare and contrast text(s). Authors should present all information using neutral language and allow readers to form their individual conclusions about the subject matter.

Authors tend to use key words such as *like, unlike, different, similar, both,* and *neither* when following a compare and contrast pattern.

Overall, organization of texts helps readers better understand an author's intent and purpose.

Drawing Conclusions

When **drawing conclusions** about texts or passages, readers should do two main things: 1) Use the information that they already know and 2) Use the information they have learned from the text or passage. Authors write with an intended purpose, and it is the readers' responsibility to understand and form logical conclusions of authors' ideas. It is important to remember that the readers' conclusions should be supported by information directly from the text. Readers cannot simply form conclusions based off of only information they already know.

There are several ways readers can draw conclusions from authors' ideas, such as note taking, text evidence, text credibility, writing a response to text, directly stated information versus implications, outlining, summarizing, and paraphrasing. Each of these are important strategies to help readers draw logical conclusions and are discussed separately.

Note Taking

When readers take notes throughout texts or passages, they are jotting down important facts or points that the author makes. **Note taking** is a useful record of information that helps readers understand the text or passage and respond to it. When taking notes, readers should keep lines brief and filled with pertinent information so that they are not rereading a large amount of text, but rather just key points, elements, or words. After readers have completed a text or passage, they can refer to their notes to help them form a conclusion about the author's ideas in the text or passage.

Text Evidence

Text evidence is the information readers find in a text or passage that supports the main idea or point(s) in a story. In turn, text evidence can help readers draw conclusions about the text or passage. The information should be taken directly from the text or passage and placed in quotation marks. Text evidence provides readers with information to support ideas about the text or passage so that they do not only rely on their own thoughts. Details should be precise, descriptive, and factual. Statistics are a great piece of text evidence because it provides readers with exact numbers and not just a generalization. For example, instead of saying "Asia has a larger population than Europe," authors could provide detailed information such as "In Asia there are over 7 billion people, whereas in Europe there are a little over 750 million." More definitive information provides better evidence to readers to help support their conclusions about texts or passages.

Text Credibility

Credible sources are important when drawing conclusions because readers need to be able to trust what they are reading. Authors should always use credible sources to help gain the trust of their readers. A text is **credible** when it is believable, and the author is objective and unbiased. If readers do not trust authors' words, they may simply dismiss the text completely. For example, if an author writes a persuasive essay, he or she is outwardly trying to sway readers' opinions to align with his or her own, providing readers with the liberty to do what they please with the text. Readers may agree or disagree with the author, which may, in turn, lead them to believe that the author is credible or not credible. Also, readers should keep in mind the source of the text. If readers review a journal about astronomy, would a more reliable source be a NASA employee or a plumber? Overall, text credibility is important

when drawing conclusions because readers want reliable sources that support the decisions they have made about the author's ideas.

Writing a Response to Text

Once readers have determined their opinions and validated the credibility of a text, they can then reflect on the text. **Writing a response to text** is one way readers can reflect on the given text or passage. When readers write their responses to the text, it is important for them to rely on the evidence within the text to support their opinions or thoughts. Supporting evidence such as facts, details, statistics, and quotes directly from the text are key pieces of information readers should reflect upon or use when writing a response to text.

Directly Stated Information Versus Implications

Engaged readers should constantly self-question while reviewing texts to help them form conclusions. Self-questioning is when readers review a paragraph, page, passage, or chapter and ask themselves, "Did I understand what I read?," "What was the main event in this section?," "Where is this taking place?," and so on. Authors can provide clues or pieces of evidence throughout a text or passage to guide readers toward a conclusion. This is why active and engaged readers should read the text or passage in its entirety before forming a definitive conclusion. If readers do not gather all the necessary pieces of evidence, then they may jump to an illogical conclusion.

At times, authors directly state conclusions while others simply imply them. Of course, it is easier if authors outwardly provide conclusions to readers because this does not leave any information open to interpretation. However, implications are things that authors do not directly state but can be assumed based off of information they provided. If authors only imply what may have happened, readers can form a menagerie of ideas for conclusions. For example, in the statement: *Once we heard the sirens, we hunkered down in the storm shelter*, the author does not directly state that there was a tornado, but clues such as "sirens" and "storm shelter" provide insight to the readers to help form that conclusion.

Outlining

An **outline** is a system used to organize writing. When reading texts, outlining is important because it helps readers organize important information in a logical pattern using Roman numerals. Usually, outlines start out with the main idea(s) and then branch out into subgroups or subsidiary thoughts or subjects. Not only do outlines provide a visual tool for readers to reflect on how events, characters, settings, or other key parts of the text or passage relate to one another, but they can also lead readers to a stronger conclusion. The sample below demonstrates what a general outline looks like.

I. Main Topic 1
    a. Subtopic 1
    b. Subtopic 2
        1. Detail 1
        2. Detail 2
II. Main Topic 2
    a. Subtopic 1
    b. Subtopic 2
        1. Detail 1
        2. Detail 2

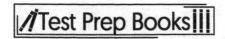

Summarizing

At the end of a text or passage, it is important to summarize what the readers read. **Summarizing** is a strategy in which readers determine what is important throughout the text or passage, shorten those ideas, and rewrite or retell it in their own words. A summary should identify the main idea of the text or passage. Important details or supportive evidence should also be accurately reported in the summary. If writers provide irrelevant details in the summary, it may cloud the greater meaning of the text or passage. When summarizing, writers should not include their opinions, quotes, or what they thought the author should have said. A clear summary provides clarity of the text or passage to the readers.

The following checklist lists items that writers should include in a summary:

Summary Checklist

- Title of the story
- Someone: Who is or are the main character(s)?
- Wanted: What did the character(s) want?
- But: What was the problem?
- So: How did the character(s) solve the problem?
- Then: How did the story end? What was the resolution?

Paraphrasing

Another strategy readers can use to help them fully comprehend a text or passage is paraphrasing. **Paraphrasing** is when readers take the author's words and put them into their own words. When readers and writers paraphrase, they need to avoid copying the text—that is plagiarism. It is also important to include as many details as possible when restating the facts. Not only will this help readers and writers recall information, but by putting the information into their own words, they demonstrate if they fully comprehend the text or passage. The example below shows an original text and how to paraphrase it.

Original Text: Fenway Park is home to the beloved Boston Red Sox. The stadium opened on April 20, 1912. The stadium currently seats over 37,000 fans, many of whom travel from all over the country to experience the iconic team and nostalgia of Fenway Park.

Paraphrased: On April 20, 1912, Fenway Park opened. Home to the Boston Red Sox, the stadium now seats over 37,000 fans. Many spectators travel to watch the Red Sox and experience the spirit of Fenway Park.

Paraphrasing, summarizing, and quoting can often cross paths with one another. The chart below shows the similarities and differences between the three strategies:

Paraphrasing	Summarizing	Quoting
Uses own words	Puts main ideas into own words	Uses words that are identical to text
References original source	References original source	Requires quotation marks
Uses own sentences	Shows important ideas of source	Uses author's words and ideas

In closing, the Paragraph Comprehension subtest will test concepts such as comprehension skills; purposes for writing; writing devices; types of passages; history and culture in relationship to literature; responding to literature; literary genres; opinions, facts, and fallacies; organization of the texts; and drawing conclusions.

Practice Questions

Directions: There are 15 questions in this practice test. Read each passage and question carefully and select the best answer.

1. When authors do not provide a clear explanation of events, what do readers often have to do instead of reading between the lines?
   a. Question
   b. Connect
   c. Infer
   d. Visualize

Use the passage below for questions 2 through 5:

Caribbean Island Destinations

Do you want to vacation at a Caribbean island destination? Who wouldn't want a tropical vacation? Visit one of the many Caribbean islands where visitors can swim in crystal blue waters, swim with dolphins, or enjoy family-friendly or adult-only resorts and activities. Every island offers a unique and picturesque vacation destination. Choose from these islands: Aruba, St. Lucia, Barbados, Anguilla, St. John, and so many more. A Caribbean island destination will be the best and most refreshing vacation ever . . . no regrets!

2. What is the topic of the passage?
   a. Caribbean island destinations
   b. Tropical vacation
   c. Resorts
   d. Activities

3. What is/are the supporting detail(s) of this passage?
   a. Cruising to the Caribbean
   b. Local events
   c. Family or adult-only resorts and activities
   d. All of the above

4. Read the following sentence, and answer the question below.
   "A Caribbean island destination will be the best and most refreshing vacation ever ... no regrets!"

What is this sentence an example of?
   a. Fact
   b. Opinion
   c. Device
   d. Fallacy

5. What is the author's purpose of this passage?
   a. Entertain readers
   b. Persuade readers
   c. Inform readers
   d. None of the above

6. Read the following sentence.
   "Cats and dogs are domesticated animals that can learn fun tricks."

In this sentence, the author is using what literary technique regarding the two animals?
   a. Comparing
   b. Contrasting
   c. Describing
   d. Transitioning

7. Which of the following is NOT an example of a fictional text?
   a. Fairy tale
   b. Autobiography
   c. Myth
   d. Fable

8. Read the following passage, and then answer the question below.

   Rain, rain, go away, come again another day. Even though the rain can put a damper on the day, it can be helpful and fun, too. For one, the rain helps plants grow. Without rain, grass, flowers, and trees would be deprived of vital nutrients they need to develop. Not only does the rain help plants grow, on days where there are brief spurts or sunshine, rainbows can appear. The rain reflects and refracts the light, creating beautiful rainbows in the sky. Finally, puddle jumping is another fun activity that can be done in or after the rain. Therefore, the rain can be helpful and fun.

What is the *cause* in this passage?
   a. Plants growing
   b. Rainbows
   c. Puddle jumping
   d. Rain

9. In fiction or drama genres, who can express dialogue?
   a. Main characters
   b. Supporting characters
   c. Narrators and characters
   d. Supporting characters and narrators

10. Read the following sentence, and answer the question below:

"If I take my socks off, my feet will get cold. But if I put socks on, my feet will get hot."

What is this sentence an example of?
    a. False analogy
    b. Circular reasoning
    c. False dichotomy
    d. Slippery slope

Use the passage below for questions 11 and 12:

Lola: The Siberian Husky

Meet Lola . . . Lola is an overly friendly Siberian husky who loves her long walks, digs holes for days, and sheds unbelievably . . . like a typical Siberian husky. Lola has to be brushed and brushed and brushed—did I mention that she has to be brushed . . . all the time! On her long walks, Lola loves making friends with new dogs and kids. A robber could break into our house, and even though they may be intimidated by Lola's wolf-like appearance, the robber would be shocked to learn that Lola would most likely greet them with kisses and a tail wag . . . she makes friends with everyone! Out of all the dogs we've ever owned, Lola is certainly one of a kind in many ways.

11. Based on the passage, what does the author imply?
    a. Siberian huskies are great pets but require a lot of time and energy.
    b. Siberian huskies are easy to take care of.
    c. Siberian huskies should not be around children.
    d. Siberian huskies are good guard dogs.

12. Because of their own experience with Siberian huskies, the author of the passage may be described as which of the following?
    a. Impartial
    b. Hasty
    c. Biased
    d. Irrational

13. Which of the following is a text that is written by a reliable source and is objective and unbiased?
    a. Narrative
    b. Persuasive
    c. Informative
    d. Credible

14. Read the following passage and answer the question below.

> "Overall, we won the championship game! Max hit a winning home run, and we all cheered as he rounded home plate. Our team hoisted the championship trophy up into the air and celebrated with joy. It was such a great game ... I will never forget this day!"

What is this passage of a story most likely to be?
   a. An outline
   b. A summary
   c. An implication
   d. A forward

15. Choose the correct way to paraphrase the following statement.

> "When renovating a home, there are several ways to save money. In order to keep a project cost effective, "Do It Yourself," otherwise known as "DIY," projects help put money back into the homeowner's pocket. For example, instead of hiring a contractor to do the demo, rent a dumpster and do the demolition. Another way to keep a home renovation cost effective is to compare prices for goods and services. Many contractors or distributors will match prices from competitors. Finally, if renovating a kitchen or bathroom, leave the layout of the plumbing and electrical the same. Once the process of moving pipes and wires is started, dollars start adding up. Overall, home renovations can be a pricey investment, but there are many ways to keep project costs down."

a. Home improvement projects can be pricey, but there are ways to keep costs down such as "Do It Yourself," or DIY; comparing prices of the materials and workers that will be used; and keeping the original floor plan for plumbing and electrical.
b. Home renovations require a lot of work. Relocating plumbing and electrical wiring can be hazardous, which is why a contractor should be hired to complete the job.
c. "Do It Yourself," otherwise known as DIY, projects are the most time- and cost-efficient way to complete home renovations. It is not necessary for homeowners to compare prices of contractors because they are their own best bet.
d. Many contractors and distributors price-match competitors' pricing for goods and services. Home improvement companies often carry the same types of product, like flooring, plumbing, or lighting. So, if completing a home renovation, it is best for homeowners to shop around for exactly what they want.

Answer Explanations

1. C: When authors do not provide a clear explanation of events, readers need to infer. An inference is made when readers draw conclusions based off of facts or evidence. Readers have to infer when authors do not provide clear answers in the text.

2. A: The topic of the passage is Caribbean island destinations. The topic of the passage can be described in a one- or two-word phrase. Remember, when paraphrasing a passage, it is important to include the topic. Paraphrasing is when one puts a passage into his or her own words.

3. C: Family or adult-only resorts and activities are supporting details in this passage. Supporting details are details that help readers better understand the main idea. They answer questions such as *who, what, where, when, why*, or *how*. In this question, cruises and local events are not discussed in the passage, whereas family and adult-only resorts and activities support the main idea.

4. B: This sentence is an opinion. An opinion is when the author states his or her judgment or thoughts on a subject. In this example, the author implies that the reader will not regret the vacation and that it may be the best and most relaxing vacation, when in fact that may not be true. Therefore, the statement is the author's opinion. A fallacy is a flawed argument of mistaken belief based on faulty reasoning.

5. B: The author of the passage is trying to persuade readers to vacation in a Caribbean island destination by providing enticing evidence and a variety of options. The passage even includes the author's opinion. Not only does the author provide many details to support his or her opinion, the author also implies that the reader would almost be "in the wrong" if he or she didn't want to visit a Caribbean island, hence, the author is trying to persuade the reader to visit a Caribbean island.

6. A: The author is comparing the two animals because they are showing how the animals are similar. If the author described the pets' differences, then they would be contrasting the animals.

7. B: Fiction is a story that is not true. An autobiography is an account of one's life written by that person, making it a true story, or nonfiction text. Fairy tales, myths, and fables are examples of fictional texts because they are made up stories.

8. D: Rain is the *cause* in this passage because it is why something happened. The effects are plants growing, rainbows, and puddle jumping.

9. C: Narrators and characters can express dialogue in fiction and drama genres. The narrator is the storyteller and can present the dialogue in either first or third person perspective. Characters play huge roles in the plot's development because they are the ones who move the storyline along.

10. B: Circular reasoning is when the reasoning is decided based upon the outcome or conclusion and then vice versa. Basically, readers can go in circles nonstop about which argument caused or had an effect on the other argument. In a false analogy, two objects or events are described as similar in a misleading or illogical way. False dichotomies, often presented as an either-or situation, are oversimplifications of arguments that neglect to address all of the possible options or sides in the argument. A slippery slope is when an author implies that something big will inevitably happen as a result of an initial smaller action.

11. A: The author implies that Siberian huskies are great pets but require a lot of time and energy. In the passage, the writer describes how huskies require lots of brushing and long walks and how they dig, making them not easy to care for. The author also describes how friendly Siberian huskies can be, even possibly greeting a robber at their own house, definitely not making them good guard dogs. Therefore, Siberian huskies are great pets but require a lot of time and energy.

12. C: The author may be biased because they show prejudice over one breed versus another in an unfair way. Impartial means fair and is essentially the opposite of biased. Hasty means quick to judge and irrational means unreasonable or illogical.

13. D: A credible text is trusted by readers, believable when the author is objective or unbiased, and written by a reliable source. Remember, text credibility is important when drawing conclusions because readers want reliable sources that support their decisions about authors' ideas.

14. B: The passage is most likely the summary of a story because it recaps the main idea, characters, what someone/something wanted, how the problem was solved, and what the result of the action was.

15. A: Paraphrasing should include the main idea and details are reworded to retell the passage. The option that best paraphrases the provided statement is: Home improvement projects can be pricey, but there are ways to keep costs down such as "Do It Yourself," or DIY; comparing prices of the materials and workers that will be used; and keeping the original floor plan for plumbing and electrical. The other options provided steer away from the topic or mention details that were not initially presented, making them incorrect.

Math Knowledge

The Scope of the Math Knowledge Section

The **Math Knowledge** section of the test involves everything included in the Arithmetic Reasoning section, as well as some additional mathematical operations and techniques. However, it is much less focused on word problems.

How to Prepare

Although this section of the test will be less focused on word problems, it is still very important to practice the types of problems in this section. As mentioned before, to really learn mathematics, it is important to practice and not just read through instructions. Approach this section the same as the Arithmetic Reasoning: first, read through the study guide here, then try the practice problems, and lastly, compare the solutions with the solutions given below. A slightly different method may be used for solving a problem, since there are sometimes multiple approaches that will work.

Numbers and Their Classification

- **Whole numbers** are the basic counting numbers.

- **Integers** are whole numbers together with the negative versions of them.

- A **factor** of an integer is a positive integer that divides it evenly. For example, 2 is a factor of 8.

- An **even number** is an integer for which 2 is a factor. If 2 is not a factor of an integer, then it is said to be an **odd number**.

- A **common factor** of multiple integers is a number that is a factor for each of them. For example, 3 is a common factor of 15 and 27. The **greatest common factor** is the largest common factor.

- A **prime number** is a whole number larger than 1 whose only factors are 1 and itself. For example, 2, 3, and 5 are prime numbers.

- A **composite number** is a whole number that is not prime. For example, 4 and 8 are not prime numbers because they have the factor 2; 21 is not a prime number because 3 and 7 are factors.

- A **multiple** of a whole number is any number that can be obtained by multiplying the whole number by another whole number. So, the first few multiples of 4 are 4, 8, 12, 16, etc.

- A **common multiple** of a set of whole numbers is a number that is a multiple of all of the numbers in the set. For example, 4 has multiples of 4, 8, 12, 16, 20, 24, etc. Whereas 6 has multiples of 6, 12, 18, 24, 30, 36, etc. In this case, 12 and 24 are examples of common multiples of both numbers. The **least common multiple** is the lowest number that is a multiple of all of the numbers in the set, which is 12, for 4 and 6.

- **Rational numbers** are numbers that can be written as a fraction whose numerator and denominator are both integers.

- A **decimal** is a number that uses a **decimal point** to show that a part of the number is less than 1. For example, $10.3 = 10 + \frac{3}{10} = 10.3$.

- The **decimal place** is how far to the right of the decimal point a digit appears. The first spot to the right of the decimal point is the *tenths* spot and indicates how many tenths are in the number. The second spot is the hundredths spot and indicates how many hundredths, and so on. The value of the numbers gets smaller the further to the right of the decimal place they are.

- **Real numbers** include rational numbers as well as any other number that can be expressed using decimals. These include things like the roots of positive numbers. Real numbers that are non-repeating and have non-terminating decimals are called **irrational**. Examples of irrational numbers are $\pi, \sqrt{7}, \sqrt{29}$, etc. Every terminating or repeating decimal number is rational. Every fraction with an integer in the numerator and denominator is rational. Fractions whose numerator and denominator are both rational numbers will also be rational.

- The number system usually used is the **decimal system**, which uses the numerals 0, 1, 2, 3, 4, 5, 6, 7, 8, and 9. Larger numbers and smaller numbers are given by writing these numerals in the appropriate place, indicating that a multiple of 10, 100, 1000, etc., is part of the number. For example, 90 indicates that there are 9 quantities of 10 in the number. Other systems are possible. Computers use a base 2 system with only 0 and 1. Babylonian mathematicians used a base 60 system. Although some other systems can also be useful, they are not needed for this test.

Operations

In this section, to simplify writing out mathematical terms, they are often abbreviated. For example, in multiplication, two numbers can be written with a dot between them: $3 \cdot 4$ rather than 3×4. Also, when working with variables, they can be written next to one another. So $xy = x \times y, 3x = 3 \times x$.

An **exponent** is written as x^y, but is read "x to the y," or "x to the power of y," and indicates (at least for whole numbers) how many times to multiply x by itself. In this expression, x is called the **base**, and y is called the **exponent**. The case where the exponent is 2 is called **squaring**, and the case where the power is 3 is called **cubing**. Again, for whole numbers, the exponent indicates how many times the base is to be multiplied by itself. So $4^3 = 4 \times 4 \times 4 = 64$. However, the exponent can be any number. The rules for working with exponents are the same, however.

A negative exponent is equivalent to switching the base from the numerator to the denominator. So $x^{-2} = \frac{1}{x^2}$, while $\frac{1}{x^{-4}} = x^4$.

The rules for working with exponents are as follows:

- $a^1 = a$
- $1^a = 1$
- $a^0 = 1$
- $a^m a^n = a^{m+n}$
- $\frac{a^m}{a^n} = a^{m-n}$
- $(a^m)^n = a^{m \times n}$

- $(ab)^m = a^m b^m$
- $\left(\frac{a}{b}\right)^m = \frac{a^m}{b^m}$

A **root** is a number that, when raised to some power, gives the number inside. It is written as $\sqrt[n]{x}$. This is called the *n*-th root of *x*, and indicates the number that, when raised to the power of *n*, gives us *x*. For this reason, it can also be written as $x^{\frac{1}{n}}$, since $(x^{\frac{1}{n}})^n = x^{\frac{1}{n}n} = x^1 = x$. If the *n* is omitted, it is assumed equal to 2. The 2nd root is called the **square root**, and the 3rd root is called the **cube root**.

A perfect square is a number whose square root is an integer. For example, 4 is a perfect square, since $4 = 2^2$. Whole numbers are either perfect squares or their square roots are irrational.

Parentheses are used to show the order in which operations are to be performed in an expression with multiple operations and terms. The rule to follow is to do the operations inside the parentheses first. However, not every operation will be marked off with parentheses.

The order in which the operations are performed is as follows:

- **Parentheses**: Do everything inside parentheses first.
- **Exponents**: First, do the operations inside the exponent, then raise the base to that exponent.
- **Multiplication**: Start on the left and work to the right.
- **Division**: Start on the left and work to the right.
- **Addition**: Start on the left and work to the right.
- **Subtraction**: Start on the left and work to the right.

Some students find it helpful to memorize this order by using a mnemonic such as **PEMDAS**: *Please Excuse My Dear Aunt Sally.*

Example: Compute $3 - 2 \times 2 + (5 - 1)^{1+1}$. The first step is to subtract inside the parentheses to get $3 - 2 \times 2 + 4^{1+1}$. The next step is to do the addition for the exponent to get $3 - 2 \times 2 + 4^2$. Next, is the exponential operation, yielding $3 - 2 \times 2 + 16$. Then, the multiplication (and division if there was any) is performed, which gives a result of $3 - 4 + 16$. The last step is to do the addition and subtraction from left to right, which gives a solution of 15.

Scientific notation is a way to write very large or very small numbers in the form $x \times 10^n$, where *x* is a number between 1 and 10. To get the number in scientific notation, the decimal point is moved *n* places to the right if *n* is positive (filling in zeroes if needed), and *n* places to the left of *n* is negative. For example, $45,000 = 4.5 \times 10^4$. As another example, $3.2 \times 10^{-5} = 0.000032$. The number with the higher exponent of base 10 is the larger number. For example, $2.9 \times 10^4 < 1.1 \times 10^5$.

A **polynomial** is an expression consisting of at least three terms, which may include constants, variables (for example, x, y, z), and exponents, involving only the four basic operations. The exponents must be non-negative integers and division by a variable is not permissible. For example, $6x^5 + 11x^4 + 6x$ is a polynomial, but $\frac{2x}{5x+1}$ is not. As with integers, a polynomial is a **factor** of a second polynomial if the second polynomial can be obtained from the first by multiplication with another polynomial. Finding the factors of a polynomial can be an involved process. Here are a few rules for factoring polynomials:

- $x^2 + 2xy + y^2 = (x + y)^2$
- $x^2 - 2xy + y^2 = (x - y)^2$
- $x^2 - y^2 = (x + y)(x - y)$

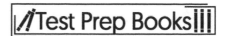

- $x^3 + y^3 = (x + y)(x^2 - xy + y^2)$
- $x^3 - y^3 = (x - y)(x^2 + xy + y^2)$
- $x^3 + 3x^2y + 3xy^2 + y^3 = (x + y)^3$
- $x^3 - 3x^2y + 3xy^2 - y^3 = (x - y)^3$

Systems of Equations

To start, a review of linear equations is needed. When given a linear equation, the equation will show two expressions containing a variable that must be equal. Thus, one example is $3x + 1 = 16$. To solve such an equation, one must remember two things. First, the final solution must equal x. Second, if two quantities are equal, one can add, subtract, multiply, or divide the same thing on both sides and end up with a true equation. In this case, 1 is subtracted from both sides, which yields a new equation, $3x = 15$. Then, both sides are divided by 3 to get $x = 5$.

A system of equations can be solved by the same kinds of considerations, except that in this case, there are multiple equations that all have to be true at the same time. This means there are some new choices for finding solutions. First, if there are two equations, the left side and the right side can be added to get a new equation (the left side of the new equation is the sum of the left sides, and the right side of the new equation will be the sum of the right sides). Second, if one equation is solved in terms of one of the variables, the expression can be substituted into the other equation. Otherwise, the approach to solving these systems is similar to solving a single equation.

A system of equations with at least one solution is called a **consistent system**. If a system has no solution, it is called an **inconsistent system**.

A **linear system** of equations with two variables and two equations is a system with variables x and y (or any other pair of variables) and equations that can be simplified to yield $ax + by = c, dx + ey = f$. There are two ways to solve such a system. The first is to solve for one variable in terms of the other and substitute it into the other equation. For example, from the first equation, $by = c - ax$, that means $y = \frac{c-ax}{b} \cdot \frac{c-ax}{b}$ can be substituted for y in the second equation. This approach is called solving by **substitution**.

The other possibility is to multiply one of the equations on both sides by some constant, and then add the result to the other equation so that it eliminates one variable. For example, given the pair $ax + by = c, dx + ey = f$, the first equation can be multiplied by $-\frac{d}{a}$. Then, the first equation becomes $-dx - \frac{db}{a}y = -\frac{cd}{a}$. Adding the equations results in the x terms cancelling, and yields an equation that only involves the variable y. This approach is called solving by **elimination**.

To illustrate the two approaches, the system of equations: $2x + 4y = 6, x + y = 2$ will be solved using both methods.

By substitution: starting with the second equation, y is subtracted from both sides. The result of this step is $x = 2 - y$. Then, $2 - y$ is substituted in for x in the first equation, with a result of $2(2 - y) + 4y = 6$. This simplifies to $4 - 2y + 4y = 6, 2y = 2, y = 1$. Then, 1 is substituted for y in $x = 2 - y$ to find the value for x: $x = 2 - 1 = 1$ or $x = 1$. So, $x = 1, y = 1$.

To solve by elimination, starting with $2x + 4y = 6, x + y = 2$: to cancel the $2x$ in the first equation, $-2x$ is placed in the second equation on the left. The second equation is then multiplied by -2 on both sides, which gives $-2x - 2y = -4$. The equations are added together:

$$2x + 4y + (-2x - 2y) = 6 - 4$$

The *x* terms cancel, and the result is $2y = 2$ or $y = 1$. Substituting this back into either of the original equations has a result of *x* = 1. So $x = 1, y = 1$.

Geometry and Angles

An **angle** describes the separation or gap between two lines meeting at a single point. It is written with the symbol $\angle$. The point where the lines or line segments meet is called the **vertex** of the angle. If the angle is formed by lines that cross one another, the vertex is the point where they cross.

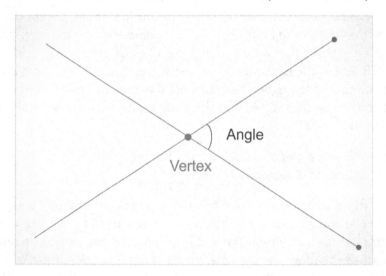

A **right angle** is 90°. An **acute angle** is an angle that is less than 90°. An **obtuse angle** is an angle that is greater than 90° but less than 180°.

An angle of 180° is called a **straight angle**. This is really when two line segments meet at a point, but go in opposite directions, so that they form a single line segment, extending in opposite directions.

A **full angle** is 360°. It is equivalent to spinning all of the way around from facing one direction back to that same direction. A full circle is considered 360°.

If the sum of two angles is 90°, the angles are **complementary**.

If the sum of two angles is 180°, the angles are **supplementary**.

When two lines intersect, the pairs of angles they form are always supplementary. The two angles marked below are supplementary:

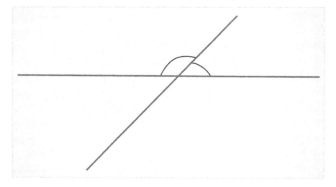

When two supplementary angles are next to one another or "adjacent" in this way, they always give rise to a straight line.

A **triangle** is a geometric shape formed by 3 line segments whose endpoints agree.

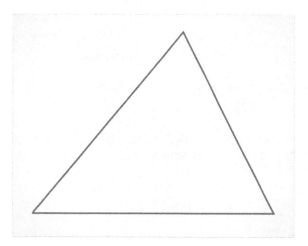

Triangle

The three angles inside the triangle are called **interior angles** and add to 180°. Triangles can be classified by the kinds of angles they have and the lengths of their sides.

An acute triangle is a triangle whose angles are all less than 90°.

If one of the angles in a triangle is 90°, then the triangle is called a **right triangle**.

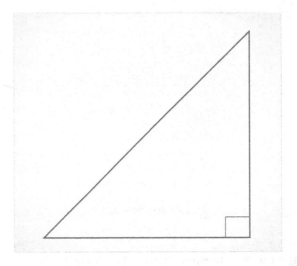

Right triangle

If one of the angles is larger than 90°, then the triangle is called **obtuse**.

An **isosceles triangle** has two sides of equal length. Equivalently, it has two angles that are the same. It can be an acute, right, or obtuse triangle.

A **scalene triangle** has three sides of different length. It also has three unequal angles. An **equilateral triangle** is a triangle whose three sides are the same length. Accordingly, its three angles are equal, and are 60°.

Consider the following triangle with the lengths of the sides labeled as A, B, C.

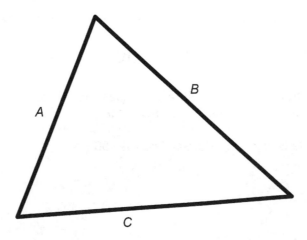

For any triangle, the **Triangle Inequality Theorem** says that the following holds true: $A + B > C, A + C > B, B + C > A$. In addition, the sum of two angles must be less than 180°.

If two triangles have angles that agree with one another, that is, the angles of the first triangle are equal to the angles of the second triangle, then the triangles are called **similar**. Similar triangles look the same, but one can be a "magnification" of the other.

Two triangles with sides that are the same length must also be similar. In this case, such triangles are called **congruent**. Congruent triangles have the same angles and lengths, even if they are rotated relative to one another.

Practice Questions

1. $\frac{14}{15} + \frac{3}{5} - \frac{1}{30} =$

    a. $\frac{19}{15}$

    b. $\frac{43}{30}$

    c. $\frac{4}{3}$

    d. $\frac{3}{2}$

2. Solve for x and y, given $3x + 2y = 8, -x + 3y = 1$.

    a. $x = 2, y = 1$
    b. $x = 1, y = 2$
    c. $x = -1, y = 6$
    d. $x = 3, y = 1$

3. $\frac{1}{2}\sqrt{16} =$

    a. 0
    b. 1
    c. 2
    d. 4

4. The factors of $2x^2 - 8$ are:

    a. $2(x - 2)(x - 2)$
    b. $2(x^2 + 4)$
    c. $2(x + 2)(x + 2)$
    d. $2(x + 2)(x - 2)$

5. Two of the interior angles of a triangle are 35° and 70°. What is the measure of the last interior angle?

    a. 60°
    b. 75°
    c. 90°
    d. 100°

6. A square field has an area of 400 square feet. What is its perimeter?

    a. 100 feet
    b. 80 feet
    c. $40\sqrt{2}$ feet
    d. 40 feet

7. $\frac{5}{3} \times \frac{7}{6} =$

    a. $\frac{35}{18}$

    b. $\frac{18}{3}$

    c. $\frac{45}{31}$

    d. $\frac{17}{6}$

8. One apple costs $2. One papaya costs $3. If Samantha spends $35 and gets 15 pieces of fruit, how many papayas did she buy?

    a. 3
    b. 4
    c. 5
    d. 6

9. If $x^2 - 6 = 30$, then one possible value for x is:

    a. -6
    b. -4
    c. 3
    d. 5

10. A cube has a side length of 6 inches. What is its volume?

    a. 6 cubic inches
    b. 36 cubic inches
    c. 144 cubic inches
    d. 216 cubic inches

11. A square has a side length of 4 inches. A triangle has a base of 2 inches and a height of 8 inches. What is the total area of the square and triangle?

    a. 24 square inches
    b. 28 square inches
    c. 32 square inches
    d. 36 square inches

12. $-\frac{1}{3}\sqrt{81} =$

    a. -9
    b. -3
    c. 0
    d. 6

13. Simplify $(2x - 3)(4x + 2)$

    a. $8x^2 - 8x - 6$
    b. $6x^2 + 8x - 5$
    c. $-4x^2 - 8x - 1$
    d. $4x^2 - 4x - 6$

14. $\frac{11}{6} - \frac{3}{8} =$

    a. $\frac{5}{4}$

    b. $\frac{51}{36}$

    c. $\frac{35}{24}$

    d. $\frac{3}{2}$

15. A triangle is to have a base 1/3 as long as its height. Its area must be 6 square feet. How long will its base be?

    a. 1 foot
    b. 1.5 feet
    c. 2 feet
    d. 2.5 feet

16. What are the zeros of the following function: $f(x) = x^3 + 4x^2 + 4x$?

    a. -2
    b. 0, -2
    c. 2
    d. 0, 2

17. What is the simplified quotient of $\frac{5x^3}{3x^2y} \div \frac{25}{3y^9}$?

    a. $\frac{125x}{9y^{10}}$

    b. $\frac{x}{5y^8}$

    c. $\frac{5}{xy^8}$

    d. $\frac{xy^8}{5}$

18. What is the solution for the following equation?

$$\frac{x^2 + x - 30}{x - 5} = 11$$

    a. $x = -6$
    b. There is no solution.
    c. $x = 16$
    d. $x = 5$

19. What is the inverse of the function $f(x) = 3x - 5$?

   a. $f^{-1}(x) = \frac{x}{3} + 5$

   b. $f^{-1}(x) = \frac{5x}{3}$

   c. $f^{-1}(x) = 3x + 5$

   d. $f^{-1}(x) = \frac{x+5}{3}$

20. Which equation is not a function?

   a. $y = |x|$

   b. $y = \sqrt{x}$

   c. $x = 3$

   d. $y = 4$

21. For the following similar triangles, what are the values of x and y (rounded to one decimal place)?

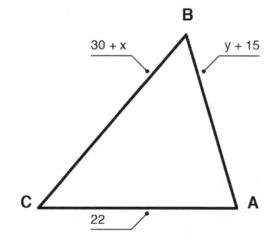
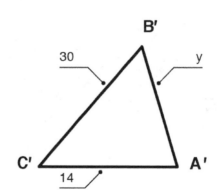

   a. $x = 16.5, y = 25.1$

   b. $x = 19.5, y = 24.1$

   c. $x = 17.1, y = 26.3$

   d. $x = 26.3, y = 17.1$

22. Which of the following correctly arranges the following numbers from the least to greatest value?

$$0.85, \frac{4}{5}, \frac{2}{3}, \frac{91}{100}$$

a. $0.85, \frac{4}{5}, \frac{2}{3}, \frac{91}{100}$

b. $\frac{4}{5}, 0.85, \frac{91}{100}, \frac{2}{3}$

c. $\frac{2}{3}, \frac{4}{5}, 0.85, \frac{91}{100}$

d. $0.85, \frac{91}{100}, \frac{4}{5}, \frac{2}{3}$

23. Simplify the following expression: $(3x + 5)(x - 8)$
   a. $3x^2 - 19x - 40$
   b. $4x - 19x - 13$
   c. $3x^2 - 19x + 40$
   d. $3x^2 + 5x - 3$

24. Simplify the following fraction:

$$\frac{\frac{5}{7}}{\frac{9}{11}}$$

a. $\frac{55}{63}$

b. $\frac{7}{1000}$

c. $\frac{13}{15}$

d. $\frac{5}{11}$

25. What is $\frac{420}{98}$ rounded to the nearest integer?
   a. 3
   b. 4
   c. 5
   d. 6

Answer Explanations

1. D: The first step is to find a common denominator of 30:

$$\frac{14}{15} = \frac{28}{30}, \frac{3}{5} = \frac{18}{30}, \frac{1}{30} = \frac{1}{30}$$

Then, the numerators should be added and subtracted, resulting in:

$$\frac{28}{30} + \frac{18}{30} - \frac{1}{30} = \frac{28 + 18 - 1}{30} = \frac{45}{30}$$

In the last step, 15 is factored out from the numerator and denominator, yielding $\frac{3}{2}$.

2. A: From the second equation, the first step is to add x to both sides and subtract 1 from both sides:

$$-x + 3y + x - 1 = 1 + x - 1$$

with the result of:

$$3y - 1 = x$$

Then, this is substituted into the first equation, yielding:

$$3(3y - 1) + 2y = 8$$

$$9y - 3 + 2y = 8$$

$$11y = 11$$

$$y = 1$$

Then, this value is plugged into:

$$3y - 1 = x, \text{ so } 3(1) - 1 = x \text{ or } x = 2, y = 1$$

3. C: The square root of 16 is ±4, so this expression simplifies to:

$$\frac{1}{2}\sqrt{16} = \frac{1}{2}(4) = 2$$

4. D: The easiest way to approach this problem is to factor out a 2 from each term:

$$2x^2 - 8 = 2(x^2 - 4)$$

The formula $x^2 - y^2 = (x + y)(x - y)$ can be used to factor:

$$x^2 - 4 = x^2 - 2^2 = (x + 2)(x - 2)$$

So:

$$2(x^2 - 4) = 2(x + 2)(x - 2)$$

5. B: The total of the interior angles of a triangle must be 180°. The sum of the first two is 105°, so the remaining is:

$$180° - 105° = 75°$$

6. B: The length of the side will be $\sqrt{400}$. The calculation is performed a bit more easily by breaking this into the product of two square roots:

$$\sqrt{400} = \sqrt{4 \times 100} = \sqrt{4} \times \sqrt{100}$$

$$2 \times 10 = 20 \text{ feet}$$

However, there are 4 sides, so the total is $20 \times 4 = 80$ feet.

7. A: To take the product of two fractions, the numerators and denominators are multiplied by their respective part in each fraction:

$$\frac{5}{3} \times \frac{7}{6} = \frac{5 \times 7}{3 \times 6} = \frac{35}{18}$$

The resultant numerator and denominator have no common factors, so this is already simplified.

8. C: Let a be the number of apples purchased, and let p be the number of papayas purchased. There is a total of 15 pieces of fruit, so one equation is:

$$a + p = 15$$

The total cost is $35, and in terms of the total apples and papayas purchased as:

$$2a + 3p = 35$$

If one multiplies the first equation by 2 on both sides, it becomes:

$$2a + 2p = 30$$

Then, this equation is subtracted from the second equation:

$$2a + 3p - (2a + 2p) = 35 - 30, p = 5$$

So, five papayas were purchased.

9. A: This equation can be solved as follows: $x^2 = 36$, so:

$$x = \pm\sqrt{36} = \pm 6$$

Only -6 shows up in the list, therefore, Choice *A* is correct.

10. D: The volume of a cube is given by cubing the length of its side.

$$6^3 = 6 \times 6 \times 6$$

216

11. A: The area of the square is the square of its side length, so $4^2 = 16$ square inches. The area of a triangle is half the base times the height, so:

$$\frac{1}{2} \times 2 \times 8 = 8 \text{ square inches}$$

The total is $16 + 8 = 24$ square inches.

12. B: Let the equation be solved as follows:

$$-\frac{1}{3}\sqrt{81} = -\frac{1}{3}(\pm 9) = \pm 3$$

Only -3 shows up in the list, therefore, Choice *B* is correct.

13. A: To solve this problem, one has to multiply each of the terms in the first parentheses and then multiply each of the terms in the second parentheses:

$$(2x - 3)(4x + 2) = 2x(4x) + 2x(2) - 3(4x) - 3(2)$$

$$8x^2 + 4x - 12x - 6 = 8x^2 - 8x - 6$$

14. C: To subtract fractions, they must have a common denominator. In this case, 24 can be used because it is the least common multiple of 6 and 8:

$$\frac{11}{6} - \frac{3}{8} = \frac{44}{24} - \frac{9}{24}$$

$$\frac{44 - 9}{24} = \frac{35}{24}$$

15. C: The formula for the area of a triangle with base b and height h is $\frac{1}{2}bh$. In this problem, the base is one-third the height, or $b = \frac{1}{3}h$ or equivalently $h = 3b$. Using the formula for a triangle, this becomes:

$$\frac{1}{2}b(3b) = \frac{3}{2}b^2$$

The problem states that this has to equal 6. So, $\frac{3}{2}b^2 = 6, b^2 = 4, b = \pm 2$. However, lengths are positive, so the base must be 2 feet long.

16. B: There are two zeros for the function $x = 0, -2$. The zeros can be found several ways, but this particular equation can be factored into $f(x) = x(x^2 + 4x + 4) = x(x + 2)(x + 2)$. By setting each factor equal to zero and solving for x, there are two solutions. On a graph, these zeros can be seen where the line crosses the x-axis.

17. D: Dividing rational expressions follows the same rule as dividing fractions. The division is changed to multiplication by the reciprocal of the second fraction. This turns the expression into:

$$\frac{5x^3}{3x^2} * \frac{3y^9}{25}$$

Multiplying across and simplifying, the final expression is $\frac{xy^8}{5}$.

18. B: The equation can be solved by factoring the numerator into $(x + 6)(x - 5)$. Since that same factor exists on top and bottom, that factor $(x - 5)$ cancels. This leaves the equation $x + 6 = 11$. Solving the equation gives the answer $x = 5$. When this value is plugged into the equation, it yields a zero in the denominator of the fraction. Since this is undefined, there is no solution.

19. D: This inverse of a function is found by switching the x and y in the equation and solving for y. In the given equation, solving for y is done by adding 5 to both sides, then dividing both sides by 3. This answer can be checked on the graph by verifying the lines are reflected over $y = x$.

20. C: The equation $x = 3$ is not a function because it does not pass the vertical line test. This test is made from the definition of a function, where each x-value must be mapped to one and only one y-value. This equation is a vertical line, so the x-value of 3 is mapped with an infinite number of y-values.

21. C: Because the triangles are similar, the lengths of the corresponding sides are proportional. Therefore:

$$\frac{30 + x}{30} = \frac{22}{14} = \frac{y + 15}{y}$$

This results in the equation $14(30 + x) = 22 \cdot 30$ which, when solved, gives $x = 17.1$. The proportion also results in the equation $14(y + 15) = 22y$ which, when solved, gives $y = 26.3$.

22. C: The first step is to depict each number using decimals. $\frac{91}{100} = 0.91$

Dividing the numerator by denominator of $\frac{4}{5}$ to convert it to a decimal yields 0.80, while $\frac{2}{3}$ becomes 0.66 recurring. Rearrange each expression in ascending order, as found in Choice *C*.

23. A: When parentheses are around two expressions, they need to be *multiplied*. In this case, separate each expression into its parts (separated by addition and subtraction) and multiply by each of the parts in the other expression. Then, add the products together.

$$(3x)(x) + (3x)(-8) + (+5)(x) + (+5)(-8)$$

$$3x^2 - 24x + 5x - 40$$

Remember that when multiplying a positive integer by a negative integer, it will remain negative. Then add $-24x + 5x$ to get the simplified expression, answer *A*.

24. A: First simplify the larger fraction by separating it into two. When dividing one fraction by another, remember to *invert* the second fraction and multiply the two as follows:

$$\frac{5}{7} \times \frac{11}{9}$$

The resulting fraction $\frac{55}{63}$ cannot be simplified further, so this is the answer to the problem.

25. B: Dividing by 98 can be approximated by dividing by 100, which would mean shifting the decimal point of the numerator to the left by 2. The result is 4.2 which rounds to 4.

Electronics Information

Electric Charge

Electricity is a form of energy, like heat or movement, that can be harnessed to perform useful work. Electrical energy results from the electric force that exists between atoms and molecules with electrical charge, which is associated with the atomic structure of those substances. Atoms contain various subatomic particles. Protons, which are in the nucleus, carry a +1 charge. Electrons, which surround the outer part of an atom in orbitals or clouds, carry a –1 charge. Net charges of atoms or molecules occur when there is an imbalance in the number of electrons and protons. A net positive charge occurs when there are more protons than electrons, while a net negative charge results when there are more electrons than protons.

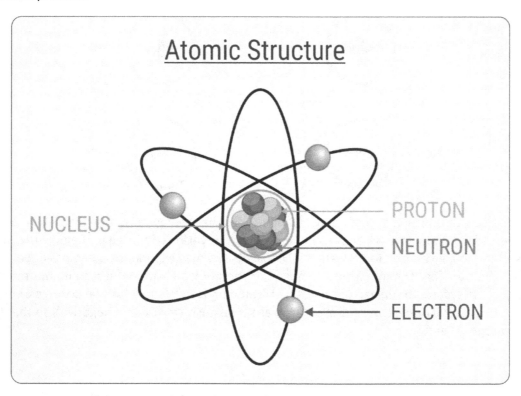

Atoms or molecules with electric charges that are the same experience a force that causes them to repel one another, while those with opposite charges attract each other. Therefore, two positive charges

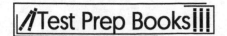

repel one another, two negative charges repel one another, but a positive and a negative charge attract one another. The unit of charge is denoted by C, the coulomb.

The Interaction of Charges

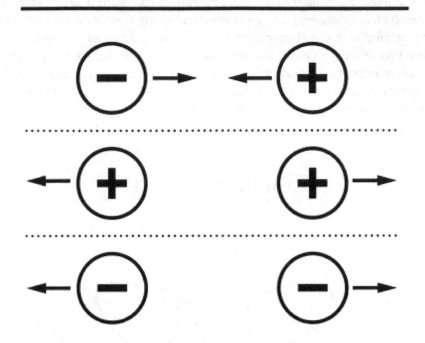

Similar to the Law of Conservation of Energy (which states that energy cannot be created or destroyed, only transferred from one form to another), there is a **conservation of charge** in the universe and in an isolated system. In a given isolated system, individual objects may experience a net loss or gain of charge with the transfer of charge from one object to another from within the system, but the overall charge within the system (or universe) cannot be created or destroyed. Individual positive or negative charges can be created or destroyed, but only in pairs (one positive with one negative), so that the net change in charge is zero.

Current

The properties of any atom vary based on the number and arrangement of electrons in the cloud surrounding the nucleus. For example, electrons in the outer orbitals of metals are relatively free to drift between atoms because they are not pulled in strongly by their atom's own nucleus. These electrons, which carry a negative charge, move through conductive materials, such as other metals, by jumping

quickly from atom to atom, creating an electrical flow. This electrical flow is energy that can be harnessed to do work.

The Flow of Electrons Creates Electrical Current

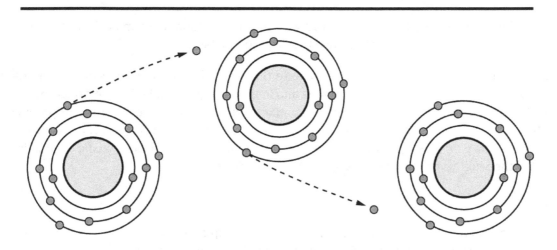

Current is the rate at which the electrical charge—or the number of electrons—flows through a conductive material. It is measured in amperes (A), with each ampere equal to approximately 6.24×10^{18} electrons per second.

Electric current carries energy much like pipes carry water. Water tanks serving an area are often elevated and have various pipes to transport the water down under ground, into each house, and then into smaller pipes leading to each sink, toilet, or shower. The purpose behind elevating the tank is to create pressure in the pipes carrying the water caused by the water above it pushing down. This pressure results in water flow, which can be equated to voltage (akin to pressure) that pushes electrons (the "water") through a circuit. With the water pipes, when the faucet is closed, water does not flow through the pipes, but water flows faster and faster the more the faucet is turned on to allow water in. The rate of water flow is analogous to current—the rate of electron flow through a circuit.

Voltage

As mentioned in the above example with the water tank, **voltage** is the push, or potential, behind electrical work. It is measured in volts (V) and can be thought of as the electromotive potential. Voltage causes current, such that if there is a closed path for electrons and a voltage, current will flow. If there is a suitable path for electrons but no voltage, or voltage in the absence of a viable path, current will cease.

A voltage source is the general term used to describe anything that can be used to generate voltage, such as a battery or generator.

Resistance

Electrical resistance, measured in ohms (Ω), is the amount of pressure inhibiting the flow of electrical current. Like friction, which slows the rate of movement, resistance dissipates energy and reduces the rate of flow or the movement of current. The amount of resistance that a given object contributes to a circuit depends on the properties of the object, particularly the material. Materials that are inherently more resistant inhibit the ease at which the electrons in the material's atoms can be displaced.

Some circuits have resistors built into them, which are specific electrical components designed to contribute a certain resistance to the circuit.

Resistance is inversely related to conductance, such that a highly conductive material has little resistance, and a material with high resistance has little conductance.

Materials vary in resistance because of the ease (or difficulty) with which electrons in the material's atoms can be displaced. The cross-sectional area and length of a given material also affect the resistance in a predictable relationship. The longer a given conductor is, the greater the resistance it provides; the greater the cross-sectional area (larger material), the less resistance there is. This relationship is quantitatively expressed as $R = \rho \cdot LA$, where ρ is the inherent resistivity of the specific conducting material, L is the length of the material, and A is the cross-sectional area.

Basic Circuits

A **circuit** is a closed loop through which current can flow. A simple circuit contains a voltage source and a resistor. The current flows from the positive side of the voltage source through the resistor to the negative side of the voltage source. Note that if the switch is open or there is some other disconnected wire or break in continuity in the circuit, there will be no electromotive force; the circuit must be a closed loop to create a net flow of electrons from the voltage source through the wires and system.

Open and Closed Circuits

A
B

Open Circuit with no electric current

Closed Circuit with electric current flowing

Ohm's Law

Ohm's Law describes the relationship between voltage, current, and resistance, which are criteria used to characterize a given circuit. The difference in electrical potential (or voltage drop) between two different points in a circuit can be calculated by multiplying the current between the two points (*I*) and the total resistance of the electrical devices in the circuit between the two points (*R*).

ΔVoltage (*V*) = current (*I*) × resistance (*R*), where *V* is voltage (in volts), *I* is current (in amperes), and *R* is resistance (in ohms).

Mathematically, the above equation can be manipulated to isolate current, which then is found to be directly proportional to the voltage drop (the electric potential difference) and inversely proportional to the total resistance.

$$I = \frac{V}{R} \ or \ R = \frac{V}{I}$$

This means that a greater battery voltage (electric potential difference) yields a greater current in the circuit, while greater resistance decreases current. Essentially, charge flows fastest when the battery voltage increases and resistance decreases.

The relationships in these equations can be understood by examining a simple circuit as a reference and then changing one variable sequentially to examine the outcome.

The Relationships Between Voltage, Resistance, and Current in a Basic Circuit

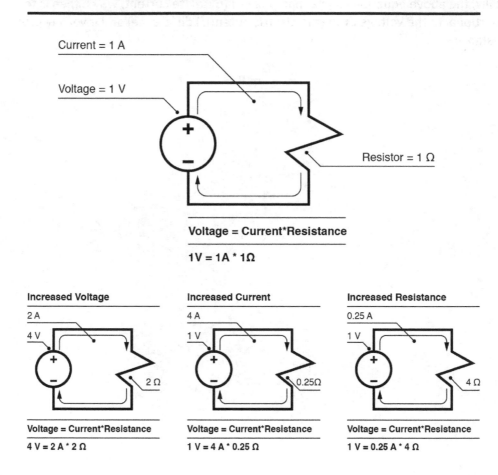

Current = 1 A

Voltage = 1 V

Resistor = 1 Ω

Voltage = Current*Resistance

1V = 1A * 1Ω

Increased Voltage

2 A

4 V

2 Ω

Voltage = Current*Resistance

4 V = 2 A * 2 Ω

Increased Current

4 A

1 V

0.25Ω

Voltage = Current*Resistance

1 V = 4 A * 0.25 Ω

Increased Resistance

0.25 A

1 V

4 Ω

Voltage = Current*Resistance

1 V = 0.25 A * 4 Ω

Alternatively, the following "Ohm's Triangle" is a useful tool to memorize the relationships governed by Ohm's Law. Test takers will need to memorize this equation for the exam. The triangle serves as a pictorial reminder and method to generate the correct relationships between voltage, current, and resistance.

Ohm's Triangle

$$V = I \times R$$

Recalling the standard Ohm's Law that $V = I \times R$ helps set up the basic triangle from which the other two equations can be visually transposed for those who find mathematically manipulating equations difficult.

Ohm's Triangle:
Voltage, Current, and Resistance Relationships

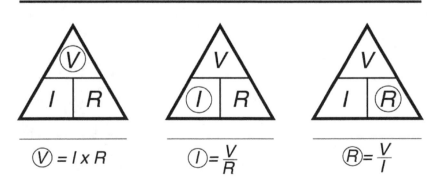

$$V = I \times R \qquad I = \frac{V}{R} \qquad R = \frac{V}{I}$$

Series and Parallel Circuits

When two or more electrical devices are powered by the same energy source (such as a battery) in a circuit, they can be connected in series or in parallel.

Series Circuits

In a **series circuit**, the electric charge passes consecutively through each device. When in series, charge passes through every light bulb. In a series circuit, there is a single pathway for electric current to flow, and all of the devices are added in succession to the same line. There are no branches coming off of this line, nor are there smaller loops within the circuit. The same electric current runs through each device, but the voltage drops as more devices or resistors are added to the string of connected devices. Essentially, the overall resistance in a series circuit increases with the addition of each resistor. This phenomenon can be observed in a series circuit composed of light bulbs. With the addition of each light bulb in succession, the brightness of each bulb decreases.

Also unique to series circuits is the interdependence of the devices connected in the series. Using the example of the light bulbs, if one of the bulbs is unscrewed or defective in some way, all bulbs in the loop (regardless of whether there is an issue with them individually or not) will go out. Simply put, for any device in a series circuit to function, all other devices in the same circuit must also be functioning.

As mentioned, the same current runs through all resistors in the series circuit, and this current must be the same across each resistor. If this was not the case, the electrons would have to build up somewhere, but current cannot be stored; the electrons going into the resistor must all go out so that the "flow in equals flow out."

While current doesn't change, voltage does drop after each resistor in such a way that the total voltage across the circuit is equal to the sum of the voltages across each device or resistor. An equivalent basic

circuit or the equivalent total resistance in the circuit can be calculated by adding the specific resistance for each resistor together.

$$R_{equiv} = R_1 + R_2 + R_3 ...$$

Equivalent Resistance in Series Circuits

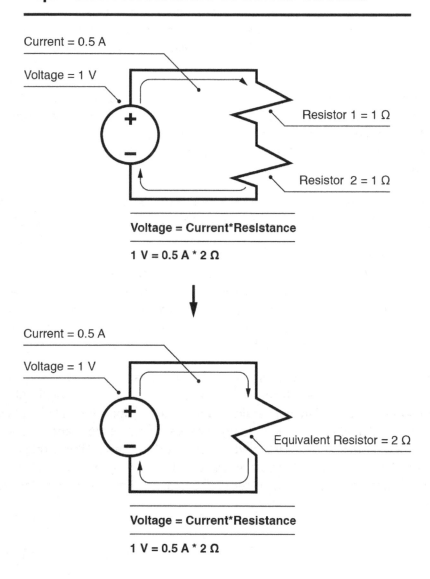

Current is the same at all locations along the circuit, whether measured at the battery, the first resistor, or after the last resistor. There is no accumulation or pileup of charge at any given location, and charge does not decrease or get used up by resistors such that there is any less charge at one point in the path compared to another.

Therefore:

$$I_{battery} = I_{resistor1} = I_{resistor2} = I_{resistor3} ...$$

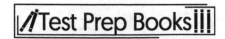

Parallel Circuits

When devices are connected in a **parallel circuit**, the charge passes through the external circuit and will only traverse one of the branches during its path back to the other terminal of the energy source. For example, if there are several light bulbs on separate branches connected in a parallel series, a single charge only passes through one of the light bulbs. Therefore, the voltage is the same across each resistor because each resistor is attached directly to the power source and the ground. In stark contrast to series circuits, as the number of devices or resistors increases in a parallel circuit, the overall current also increases. Somewhat counterintuitively, the addition of more resistors in a separate branch of the circuit actually decreases the overall resistance in the parallel circuit, so the current increases.

Another difference between parallel and series circuits is that the removal or malfunctioning of one device in the parallel circuit does not affect the overall current or the current in the other branches. This feature makes parallel circuits advantageous for many applications, such as home wiring, to prevent a total failure of multiple devices and with a defect in a single device. In summary, in series circuits, the current stays the same, but the voltage drops, while in parallel circuits, the current increases, and the voltage stays constant.

A useful analogy to differentiate between series and parallel circuits as well as to understand that adding resistors to a parallel circuit actually increases current is to equate each series to tollbooths on highways. Bottlenecks can occur at points along the highway where cars must pay tolls. The flow rate of the cars is reduced significantly in these areas. Therefore, the tollbooths are analogous to resistors, and the rate of car travel is like the electric current. To remedy the bottlenecks that occur, the transportation authority may decide to add three more tollbooths, which is essentially adding more resistors. If the tollbooths were added in series, each car would have to stop and pay a toll at every tollbooth in a consecutive manner. If the toll was originally $.80, the addition of the three new tollbooths would result in a car stopping at all four tollbooths and paying $.20 per stop.

This would have the obvious effect of increasing the jam (total resistance), and traffic flow rate (electric current) would stay the same because still only one car at a time could pass through. Alternatively, if the three additional tollbooths were added in a parallel fashion, each tollbooth would be placed on a separate branch of the road, and cars would select one of the four possible branches and stop at only the one tollbooth in their branch. In contrast to adding tollbooths in series, this arrangement would increase traffic flow and improve the initial bottleneck issue by creating more pathways by which cars (or charge) can flow. Essentially, the electric current increases over baseline in parallel circuits and decreases in series.

Here is an example image:

Altering the Flow Rate on a Tollway

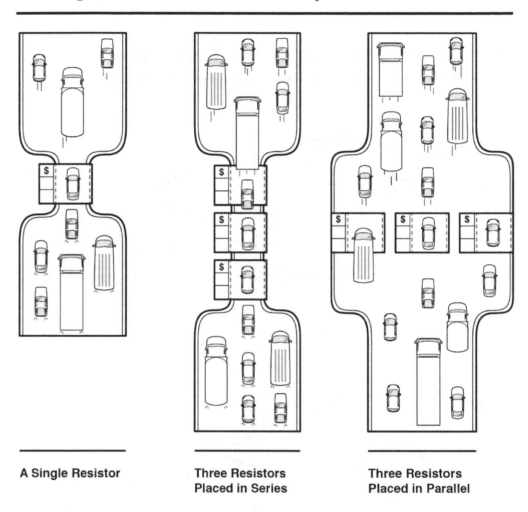

A Single Resistor   **Three Resistors Placed in Series**   **Three Resistors Placed in Parallel**

Like with series circuits, resistors in parallel can also be reduced to an equivalent circuit, but not by simply adding the resistances. The equivalent resistance (R_{eq}) is instead found by solving Ohm's Law for the current through each resistor, setting this value equal to the total current (I_t), and remembering that the voltages are all identical. Essentially, this yields an equation that shows that the inverse of the equivalent resistance of parallel resistors is equal to the sum of the inverses of the resistance of each leg of the parallel circuit.

Mathematically, that means:

$$I_t = \frac{V}{R_{eq}} = \frac{V}{R_1} + \frac{V}{R_2} \text{ or } \frac{1}{R_{eq}} = \frac{1}{R_1} + \frac{1}{R_2} \text{ so } R_{eq} = \frac{1}{\frac{1}{R_1} + \frac{1}{R_2}}$$

Equivalent Resistance in Parallel Circuits

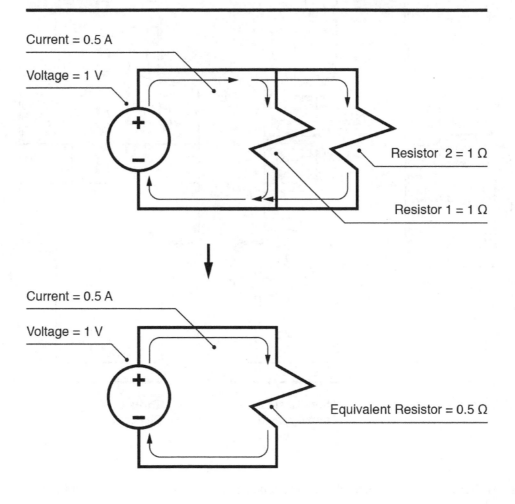

$$R_{equiv} = \frac{1}{\frac{1}{1\,\Omega} + \frac{1}{1\,\Omega}} = 0.5\,\Omega$$

As mentioned, the total resistance decreases as additional resistors are added to the parallel series, and the equivalent resistance is always lower than that of the smallest resistor.

Electrical Power

Electrical power in a circuit refers to the energy produced or absorbed, and it is expressed in watts (W). In a given circuit, certain components such as light bulbs consume electrical power and convert it to

light/heat, while other components, for example, the battery, produce power. Power, in watts, is equal to the current from a voltage source in amperes multiplied by the voltage of that source:

$$\text{Power }(W) = \text{current }(I) \times \text{voltage }(V)$$

Using Ohm's Law and substitution, this relation can also be written in the following two ways:

$$W = I^2 R$$

$$W = \frac{V^2}{R}$$

Of note, sometimes power ratings (the rate that electric power is converted to a different form of usable energy [such as light or motion]) on electric motors or other devices is expressed in horsepower, or hp, where 1 hp = 746 W.

AC Versus DC

The circuits previously described are **direct current (DC) circuits**, which are circuits wherein the voltage source has a constant value, and the current flows unidirectionally. This type of circuit is usually used for the wiring on ships, airplanes, and electronic devices like cell phones and televisions. Batteries produce DC, and devices called rectifiers convert **alternating current (AC)** to DC. In contrast, voltage alternates over time in AC circuits, and current flow can switch directions. AC electricity is used in most land-based, heavy machinery and powers houses and buildings because large amounts of AC electricity can be transmitted over long distances with significantly less loss of power than with DC electricity. In the United States, AC electricity is generated (via alternators) at a frequency of 60 cycles per second with 120- and 240-volt service available to users via a three-wire distribution system.

Capacitors

Capacitors are devices that store electric charge and resist changes in voltage. They are made from two parallel plates made of conductive surfaces that are separated by a space or an insulating material such as ceramic or Teflon. Capacitors filter signals by blocking DC signals but permitting AC ones.

Capacitors

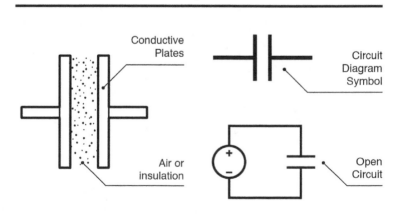

Inductors

Inductors are coils of wire that develop a magnetic field in the presence of electrical flow. This magnetic field resists changes in the magnitude of changes in current in AC circuits. They can function as filters because they block AC signals but not DC, and they can separate signals of varying frequencies.

Inductors

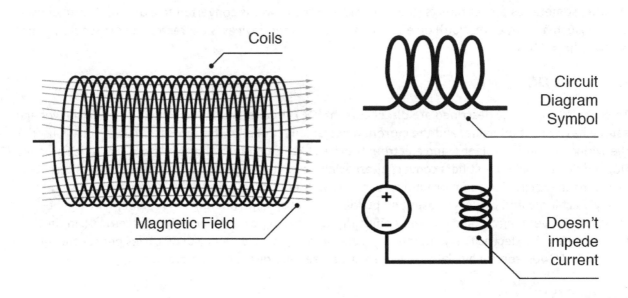

Electrical Diagrams

Test takers should be familiar with reading **block diagrams**, which depict electrical circuits with various symbols. The following figure includes frequently encountered symbols:

Common Circuit Diagram Symbols

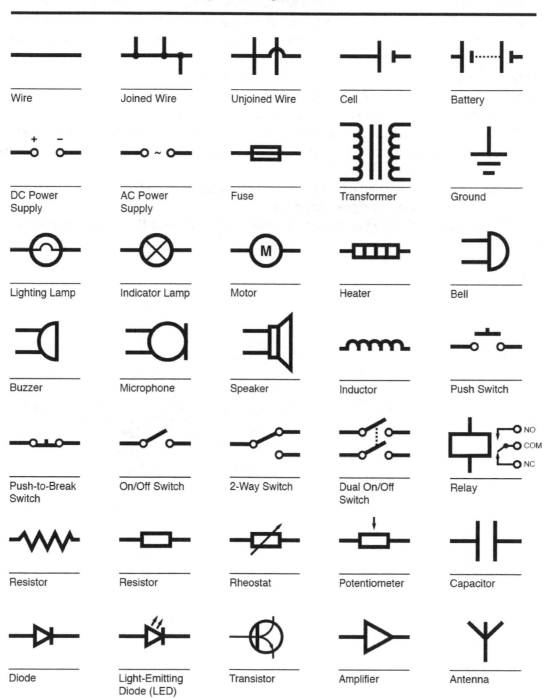

Here are some basic definitions for important terms:

- **Cell**: The electronic unit that supplies current; batteries contain multiple cells.
- **Fuse**: A safety device that protects against current that surpasses a specified value by opening the circuit to prevent current flow.
- **Transformer**: A device that increases or decreases AC voltage by transferring energy via magnetic field induction between two coils of wire that are not electrically connected.
- **Ground**: The electrical connection to the earth, which enables the ground to serve as a zero-point reference for voltage against which other voltages in a circuit can be measured.
- **Transducer**: A device, such as a motor or lamp, that converts energy from one form into a different form.
- **Relay**: A type of switch that can open or close a circuit electromagnetically or electrically.
- **Rheostat**: A two-terminal variable resistor that controls current (such as lamp brightness).
- **Potentiometer**: A three-terminal variable resistor that controls voltage.
- **Diode**: The electrical equivalent of a valve; it permits electricity to flow in only one direction, indicated by the arrow in diagrams. Light-emitting diodes (LEDs) illuminate when current passes through.
- **Transistor**: A type of semiconductor (not quite a conductor yet not fully an insulator) that amplifies or switches electric signals or power.
- **Amplifier**: An electrical circuit contained in a device that increases the voltage, power, or current of a signal.

Practice Questions

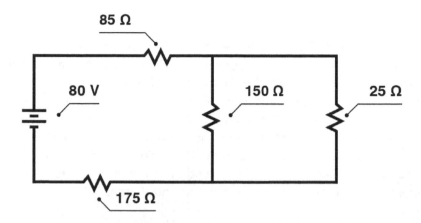

1. What is the total resistance in the circuit above?
    a. 85 ohms
    b. 435 ohms
    c. 278.75 ohms
    d. 260 ohms

2. What is the total current leaving the battery?
    a. 22,300 Amps
    b. 0.29 Amps
    c. 80 volts
    d. 80 Amps

3. What is the voltage lost at the 25-ohm resistor?
    a. 0.22 amps
    b. 0 amps
    c. 7.25 amps
    d. 5.4 volts

4. A 150-watt waffle iron is plugged into a 200-volt circuit at a diner. The waffle maker operates for 240 minutes. If the cost of energy is $.10 per kilowatt-hour, how much does it cost to use the waffle maker for the 240 minutes?
    a. $60
    b. $3.60
    c. $6.00
    d. $.60

5. The energy from electricity results from which of the following?
   a. The atomic structure of matter
   b. The ability to do work
   c. The neutrons in an atom
   d. Conductive materials like metals

6. Which of the following correctly lists the subatomic particles in an atom?
   a. Protons carry no charge in the nucleus, neutrons carry a negative charge in the nucleus, and electrons surround the nucleus in a cloud and carry a positive charge.
   b. Protons carry a positive charge in the nucleus, neutrons have no charge in the nucleus, and electrons have a negative charge and are found in a cloud surrounding the nucleus.
   c. Protons carry a positive charge in the nucleus, electrons have a negative charge in the nucleus, and neutrons have a negative charge and surround the nucleus in a cloud.
   d. Protons carry a positive charge, and they surround the nucleus in a cloud. The nucleus contains negatively charged electrons and neutrons, which carry no charge.

7. What property of metals lends them to be conductive materials?
   a. The protons in the outer shells are free to drift between atoms.
   b. The electrons in the outer shells are free to drift between atoms.
   c. The electrons in the outer shell are pulled in tightly by the nucleus so they do not drift away to surrounding atoms.
   d. The neutrons in the outer shell are pulled in tightly by the nucleus so they do not drift away to surrounding atoms.

8. Which of the following best describes the function of and symbol for a diode?

   a. It is a type of variable resistor that controls voltage, and it is represented as:

   b. It is a type of variable resistor that controls voltage, and it is represented as:

   c. It ensures that current flows one way only, and it is represented as:

   d. It ensures that current flows one way only, and it is represented as:

9. In the following circuit, what is the total resistance across the two terminals (A and B)?

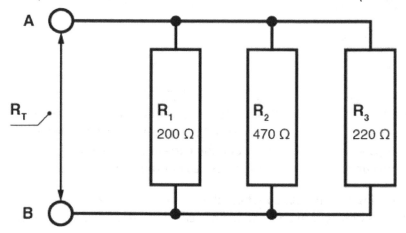

a. 85.67 ohms
b. 790 ohms
c. 0.0117 ohms
d. 200 ohms

10. Which of the following statements is true?
a. Conductors have high resistance because the electrons are easily detached from the atoms.
b. Conductors have high resistance because the electrons are not easily detached from the atoms.
c. Conductors have low resistance because the electrons are easily detached from the atoms.
d. Conductors have low resistance because the electrons are not easily detached from the atoms.

11. Which of the following statements are true regarding insulators?
a. Insulators provide more resistance than conductors; copper is a good insulator.
b. Insulators provide less resistance than conductors; copper is a good insulator.
c. Insulators provide more resistance than conductors; plastic is a good insulator.
d. Insulators provide less resistance than conductors; plastic is a good insulator.

12. Which of the following equations is not an accurate representation of Ohm's Law?
a. $V = I \times R$
b. $R = \frac{V}{I}$
c. $I = \frac{V}{R}$
d. $V = \frac{R}{I}$

13. Electric power can be measured in which of the following units?
a. Volts
b. Amperes
c. Ohms
d. Watts

14. How much energy is needed to run a 6000-watt air conditioner for 3 hours?
    a. 18 kilowatts
    b. 18 kilowatt-hours
    c. 18,000 kilowatt-hours
    d. 18,000 kilowatts

15. Which of the following is true regarding series circuits?
    a. The voltage drops across each resistor, but the current is the same in all of them.
    b. The voltage is the same across each resistor, but the current drops after each of them.
    c. The voltage drops across each resistor, and the current drops in each of them.
    d. The voltage and current are the same across each of them in the series.

16. Which of the following statements about parallel circuits is FALSE?
    a. If the resistance is different in each load, more current passes through the load with the lower resistance.
    b. If the resistance is different in each load, more current passes through the one with higher resistance.
    c. If the resistance is the same in both loads, then the same amount of current passes through each one.
    d. The voltage is the same across each resistor.

17. Which of the following is not a correct formula to determine electric power in watts?
    a. $W = V \times I$
    b. $W = I^2 \times R$
    c. $W = V^2/R$
    d. $W = I^2 \times V$

18. Which of the following statements about circuits is NOT true?
    a. The wiring on land-based, heavy machinery uses DC electricity.
    b. The voltage source for DC circuits has a constant value.
    c. AC electricity can be transmitted over long distances with much less loss of power than DC electricity.
    d. In the United States, AC electricity is generated at a frequency of 60 cycles per second with 120- or 240-volt service.

19. All EXCEPT which of the following statements are true regarding capacitors?
    a. It is a set of parallel plates separated by a non-conducting material.
    b. They don't stop ACs, but they do stop DCs.
    c. The electric charges built up on either side of the capacitor resist change equal to the voltage of the power source.
    d. The coils in the electric motors of AC equipment are examples of capacitors.

20. If a circuit has a 9-volt battery and a 3-ohm resistor, what is the power output of the battery?
    a. 27 watts
    b. 243 watts
    c. 81 watts
    d. 6 watts

Answer Explanations

1. C: Total resistance in a circuit that has both series and parallel elements is calculated by the following equations:

$$\text{Total resistance} = 85\ \text{ohms} + 175\ \text{ohms} + 1/(1/150\ \text{ohms} + 1/25\ \text{ohms})$$

$$\text{Total resistance} = 260\ \text{ohms} + 18.75\ \text{ohms}$$

$$\text{Total resistance} = 278.75\ \text{ohms}$$

2. B: The total resistance for the circuit was found in the previous problem (278.75 ohms), so this value is plugged into the equation as well as the voltage from the battery source (80 volts).

$$I = V/R = 80\ V/\ 278.75\ \text{ohms} = 0.29\ \text{amps}$$

Current is measured in amps, not volts, so Choice C is incorrect. The other choices do not manipulate the equation appropriately.

3. A: Voltage lost is expressed in amps because it is referring to a reduction in current, so Choice D, 5.4 volts, is incorrect. Voltage will be lost at the 25-ohm resistor because this resistor is in a parallel circuit, so Choice B, 0 amps, is incorrect. Again, the voltage lost can be calculated by substituting known values into iterations of Ohm's Law.

$V = I \times R = (0.29\ \text{amps}) \times (18.75\ \text{ohms}) = 5.4\ \text{volts}$. Then, this is substituted into the following equation:

$I = V/R = 5.4\ V/25\ \text{ohms} = 0.22\ \text{amps}$

4. D: The equation to determine the cost is cost = power (in kilowatt-hours) × time (in hours) × rate (in cents per kilowatt-hour). The important step is ensuring that all values are inserted in the equation with the correct units, so some conversions need to take place. Power is given as 150 watts, which equals 0.15 kilowatt-hours (150/1000). The time of 240 minutes is divided by 60 minutes/hour to yield 4 hours. Therefore:

$$\text{cost} = (0.15\ \text{kW}) \times (4\ \text{hr.}) \times (10\ \text{cents/kWh}) = 6\ \text{cents}/100 = \$.06$$

Choice A did not convert the power into kilowatt hours but left it in watts. Choice B did not convert the time from minutes into hours. Choice C did not correctly convert the cost into dollars.

5. A: The physical structure of the atoms that compose matter lends itself to the production of electricity. The arrangement of the subatomic particles and the associated charges—mainly the negatively charged electrons in the cloud—are associated with the ability to create an electric current, which can be harnessed to do work.

6. B: The structure of atoms is as follows: The central nucleus contains protons and neutrons. Protons have a positive charge (+1), and neutrons are neutral, so they do not carry a charge. Around the nucleus is a cloud, which contains electrons in rings called orbitals. Electrons are negatively charged (–1), and those in the outermost ring—farthest from the nucleus—tend to be easier to remove from the atom.

7. B: Physical properties of atoms are influenced to some degree by the number and arrangement of electrons in the shells around the nucleus. The electrons in the outer shells in metals are free to drift between atoms. This movement potential allows them to jump between atoms, creating an electric current or flow, when subjected to an external force such as a magnetic field.

8. C: Diodes function in electrical circuit much like valves do with fluid dynamics or in veins. They prevent current from flowing in the unintended direction and instead only permit flow in one direction. The symbol is:

$$\dashv\!\!\triangleright\!\vdash$$

The arrow points in the direction of permitted current flow. Choices *A* and *B* described potentiometers, and Choice *D* showed the symbol for the potentiometer.

9. A:

$$\frac{1}{R_{eq}} = \frac{1}{R_1} + \frac{1}{R_2} + \frac{1}{R_3} + \cdots$$

So, in this circuit,

$$\frac{1}{R_{eq}} = \frac{1}{220} + \frac{1}{200} + \frac{1}{470}$$

$$\frac{1}{R_{eq}} = \frac{1}{0.0117} \quad so \ R_{eq} = 85.67 \ ohms$$

Choice *B* added the resistances together as is done in a series circuit, but the given circuit is a parallel circuit. Choice *C* missed the final step that the resistance found is the inverse of R_{eq}. Choice *D* simply chose the lowest resistance of the given resistors, but this is not how to calculate the equivalent resistance in parallel circuits.

10. C: Electric conductors tend to be metals, such as silver, copper, and aluminum. They "conduct" electricity, which means that they help the electric current flow easily. This is largely because metals have low resistance because the electrons are easily detached from these atoms, so they are then free to jump from atom to atom and create a current. In contrast, insulators (such as wood, plastic, or rubber) have high resistance but low conductivity. Conductivity and resistance are opposite properties; a given single material cannot be highly conductive and highly resistant at the same time.

11. C: Insulators provide more resistance than conductors. As mentioned, conductors have little resistance. Insulators, such as plastic or rubber, in contrast, have high resistance. Essentially, they slow the flow of electric current much like friction deters mechanical motion.

12. D: Ohm's Law describes the relation between voltage and amperage, where voltage is a measure of electromotive potential much like potential energy of motion in kinetics. Amperage measures the electric current or the flow rate of 1 coulomb of electrons in a second. The basic law is $V = I \times R$, but this can be manipulated in the following ways:

$$I = \frac{V}{R} \quad or \quad V = I \times R \quad or \quad R = \frac{V}{I}$$

13. D: The difference in potential between two separate points is a measure of volts. Amperes indicate the number of electrons that pass a specific point per second. Ohms measure the resistance to flow (essentially electrical friction). Watts are a measure of power or the rate at which electrical energy is used or converted into another type of energy. Watts are often converted to kilowatts or, when expressing power rates, kilowatt-hours.

14. B: The amount of energy needed to power a certain device is expressed in kilowatt-hours. Therefore, Choices *A* and *D* are incorrect. The amount of energy needed is the product of the power in kilowatts multiplied by the time in hours. The air conditioner is 6000 watts or 6 kilowatts multiplied by 3 hours is 18 kilowatt-hours. Choice *C* did not properly convert the power from watts into kilowatts.

15. B: When devices are connected in a series circuit, the voltage is the same across each device (or resistor) in the loop, but the current drops after each device. Recall that for lightbulbs connected in series, this is observed by a decrease in the brightness of each successive bulb. This is one of the disadvantages of a series circuit. In contrast, the current increases with the addition of each resistor in parallel circuits.

16. B: In a parallel circuit, more charges will seek the path with least resistance. Because of this, more current passes through circuit paths with lower resistance. If the resistance is the same in both paths, then the same amount of current passes through each resistor. An important feature of parallel circuits is that the voltage is the same across each resistor; it does not drop as it does in series circuits. In series circuits, voltage drops after each resistor, and current is constant, and in parallel circuits, voltage stays the same after each resistor, and current increases. The total current can be calculated as the sum of the electric current in each branch of the parallel circuit.

17. D: Choice *A* depicts the basic equation for power, *W*, expressed in watts is $W = V \times I$, where *V* is the voltage in volts and *I* is the current in amps. Using Ohm's Law and substitution, $V = I \times R$ can replace the *V* so that $W = (I \times R) \times I$ or $I^2 \times R$, which was Choice *B*. Ohm's Law also holds that $I = V/R$, so substituting this equation for the *I* in the basic power equation yields $W = V \times (V/R)$ or $W = V^2/R$. Choice *D* is not a correct substitution and use of Ohm's Law and the power equation.

18. A: Most land-based, heavy machinery use AC electricity, not DC, because large amounts of electricity can be transmitted over long distances with AC while incurring significantly less loss of power than with DC electricity. DC circuits have voltage sources that provide a constant voltage, and the wiring on most ships, airplanes, and electronic circuits uses DC electricity.

19. D: The coils in the electric motors of AC equipment serve as inductors (not capacitors), which are typically coils of conducting wire in which a magnetic field is created by the electric current. Examples of capacitors are ceramic, Teflon, and air because these are non-conducting materials.

20. A: Electrical power is essentially the energy output over a given time. It is found by multiplying the current (in amps) from a voltage source by the voltage of that source. So, $P = IV$. Ohm's Law states that the voltage equals current in the circuit multiplied by the resistance ($V = IR$). Therefore, the power equation can be written two different ways, by substituting known relations from Ohm's Law:

$$P = I^2R \text{ or } P = \frac{V^2}{R} \qquad So\ Power = \frac{V^2}{R} = \frac{9^2}{3} = 27\ watts$$

Auto and Shop Information

The following pages will contain information that informs the reader about the most essential tools and techniques necessary for working on American-made vehicles and American-manufactured combustion engine automobiles. The information contained in this manual is crucial to effectuating important tasks required for servicing automobiles.

All engines and vehicle systems require maintenance; common everyday tools and task-specific tools are required to complete vehicle maintenance. An array of screw drivers, wrenches, hammers, and other task-specific tools must be available in order to successfully fix the various systems of a vehicle.

Types of Automotive Tools

There are several tools that are indispensable to the auto mechanic:

- **Wrenches**: A mechanic cannot do without an assortment of wrenches: The 9/16" and ½" fixed open wrench are among the most common and necessary tools for the vehicle mechanic. Most nuts and bolts in an American-made vehicle have numerous nine-sixteenths and one-half inch nuts and bolts. These wrenches are important for taking apart connections and housings that are located inside the vehicle. A multitude of wrenches of different sizes are required to fix the many different types of vehicles that exist in the world today. Variable wrenches that can adjust in size are useful in cases where the available fixed size wrenches are not of the specific size needed for an oddly-sized nut or bolt.

  A complete ratchet set is also a necessary kit for the mechanic. A ratchet tool has a handle and a right angle circular notched head that fits over the nut. Different sized ratchets attach to the right-angle at the end of the tool's handle.

- **Vice grips**: Medium-sized vice grips lock into place in order to hold onto something, which enables the mechanic to grip items with a grip-locking feature. Vice grips often work as another pair of hands to hold something while the mechanic implements repairs. Various sizes of vice grips are important tools in a mechanic's arsenal because the mechanic can position and hold items with the assistance of the vice grips, which lock the item being held into place.

- **Channel lock pliers**: The jaws on these pliers are positioned at a horizontal right-angle, which enables the mechanic to grab onto a nut or bolt that would be hard to turn without the right-angle jaw feature of the channel lock pliers. Channel lock pliers are also adjustable. The gap between the jaws is variable due to a tongue and groove sizing system that lets the user adjust the jaw gap by the tongue and groove adjustment feature located below the jaws. Every mechanic should have at least one channel pliers. Some channel pliers have rubber handles that protect the mechanic from electrical shock when working near wires that contain a live electric current.

- **Flat head screwdrivers**: One large screwdriver, preferably about 10 inches long, is an important tool that can do double duty as a small pry bar. The large screwdriver is indispensable for turning those larger screws the mechanic will encounter. A smaller flat head screwdriver is good for the more diminutive slotted screws. The 5-inch screwdriver is also useful for prying things up

when a smaller tool is required. Various sized flat head screwdrivers round out the mechanic's toolbox so that he or she can perform many different types of vehicle repairs.

- **Phillips-head screwdrivers**: the four slotted "Phillips" screws are encountered throughout a vehicle and various sized Phillips-head screwdrivers are important for the mechanic to have on hand to effectuate repairs.

Automotive Systems

Important Automotive Components

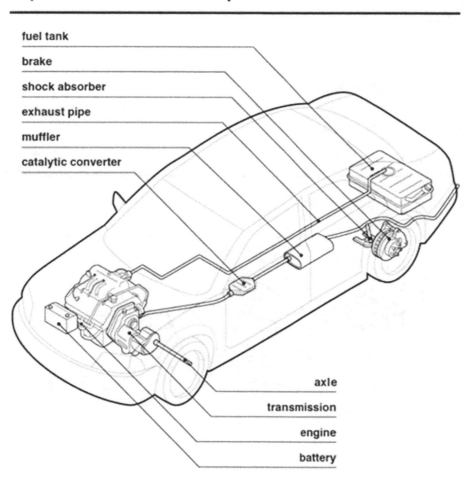

There are many types, makes, and models of vehicles and just as many different companies that manufacture vehicles, but the focus here is on regular gas combustion engine vehicles. Electrically-powered vehicles and self-driving vehicles are starting to become more commonplace, but the electrical and artificial intelligence technology is still in its relative infancy. The main source of power for vehicles around the world is still the gasoline combustion engine.

Primary systems of the vehicle must be maintained and repaired. The primary systems of a vehicle include, but are not limited to:

- Engine System
- Ignition System
- Electrical System
- Drive System
- Lubrication System
- Cooling System
- Exhaust System
- Brake System
- Suspension System
- Steering System

Each system will be reviewed briefly.

The Engine System

Engine Sketch

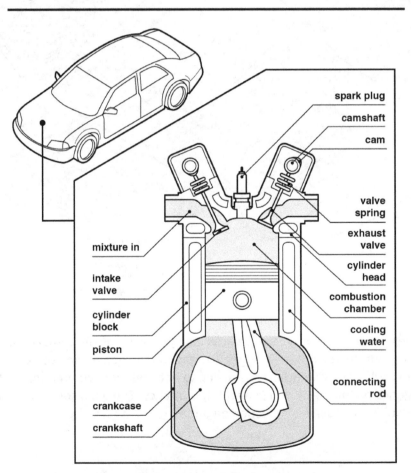

Engine Systems and Components

Primary engine systems for a gasoline combustion-based motor include:

- **Engine block**: the main frame or central structure of the engine upon which every other engine piece rests or functions within the block

- **Pistons**: cylindrical objects that rest like hats on movable metal arms that take in and compress gas within the piston chamber

- **Spark plugs**: electrical components that create a spark that ignites gas within the piston chamber

- **Rocker arms**: a small hinged metal arm that is positioned horizontally over the piston chamber to control the intake valve and exhaust valve of the piston chamber

- **Cams**: steel nodes along the camshaft that are machined to operate the rocker shafts or rocker arms

- **Valves**: an intake valve and an exhaust valve is positioned over every cylinder chamber

- **Fuel injectors**: nozzles that inject fuel into the piston chamber

- **Carburetors**: air intake components that intake air to feed to the engines' combustion process

- **Electrical wires**: numerous and increasingly complex wires that carry electrical current to all auto systems are under the hood of all modern cars

- **Starters**: a starter module and a starter motor starts the combustion process in a gas engine

- **Gaskets**: fitted pieces of thin material, formed to fit exactly over an orifice in order to keep gas or liquid from leaking out

Using the proper tools for repairs is important because all of the engine's components can be repaired or replaced but using the proper tool for the task is essential.

The **power system** of the automobile is the engine. For this reason, it is perhaps the most essential system. The engine is responsible for the operation of many other systems in the vehicle. Without the power of the engine, many systems—such as the drive system or the cooling system—would not operate.

Many different types of repairs to an engine might have to be completed. One common repair that is needed is to fix the valve cover gaskets. A **valve cover gasket** is a rubber-like fitted piece that fits precisely to the dimensions of the valve cover and seals off a section of the engine over the valves to prevent oil from spewing out. The gaskets also prevent air from unintentionally entering the engine. The valves of the pistons are located directly above the pistons. These valves are covered by metal rectangular castings that protect the spark plugs. The heat produced at the top of the engine can degrade the valve cover gaskets or "seals," which causes oil to spurt out from the gaps in the worn gaskets, necessitating replacement. Variable wrenches are helpful to loosen the nuts holding down the valve cover gaskets.

The engine works in concert with many other systems in the vehicle and the engine also provides power to certain other vehicle systems, such as "power steering" and the electrical system. All other systems and components work together with the engine's power; such as braking components, steering components, shocks, exhaust mufflers, and exhaust pipes. There are many job-specific tools made to handle the repairs of these vehicle systems.

The Ignition System
Ignition Systems and Components
The car's **ignition key**, when turned in traditional cars or pushed in some newer vehicles, completes an electrical connection from the car's battery and sends current to the ignition module that controls the electrical ignition motor. The electrical **ignition motor,** or starter motor, turns the crankshaft, which then initiates the up and down motion of the cylinders while the valves above the cylinders inject gas into the piston chambers. The **spark plugs** ignite the gas and create combustion in the cylinders.

The **ignition** is the system that can be the most problematic because so many things must happen simultaneously and in synchronization for the combustion process to start within the piston chambers. Just one wire connected improperly or an insufficiently charged vehicular battery can result in non-ignition.

Combustion inside the engine cylinders is a self-perpetuating power system, as long as gas continually feeds into the cylinders. When the key is turned, it completes an electrical connection between the car's battery and the starter motor. The starter motor is electrically driven by the battery and turns the crankshaft. When the crankshaft turns, the pistons are powered into motion and travel up and down in sealed cylindrical chambers. When the piston rises, it compresses the vaporized gasoline that has been introduced into the piston chamber. When the gas is compressed by the pistons' "up-stroke," a sparkplug "fires" or sparks, and ignites the compressed gas inside the piston chamber. When the gas inside the piston chamber explodes, the resulting explosive force drives the piston downward. This downward piston force is transferred to the crankshaft of the engine, which then transfers that kinetic energy to the transmission system and drive system of the vehicle.

The **alternator** is an electrical generator powered by the gas engine; it is an important main unit in the vehicle's electrical system, which produces electricity so the sparkplugs will keep firing for the duration of the vehicle's operation.

If the sparkplugs are fouled, a buildup of carbon occurs between the small gap of metal that makes the spark, which hinders a spark from forming. This carbon buildup between the spark gap located at the end of the sparkplug occurs from extensive engine operation, the residual buildup of carbon from the incendiary action of the spark, and the aftermath of exploding gases. The resulting carbon deposits prevent the sparkplug from "firing" or sparking, which normally enables the cylinder gases to fire. The engine will not work properly or start if the carbon buildup between the sparkplug gap prevents the sparking from taking place. The spark ignites the gases in the engine's piston chamber. Without the gas explosion in the piston chamber, the engine will not "crank" or start. **Sparkplug fowling** is a common malady that affects the ignition system of a vehicle.

Thankfully, it is a relatively easy fix for the mechanic with the right tools and knowledge. The first step is to take the sparkplug out by using a specific tool called a sparkplug wrench. This task-specific wrench has a small rubber integral part in the jaws that pulls the sparkplug out of its placement. The mechanic must clean the sparkplug contact with coarse sand paper.

The **coil wire** provides the electricity to the individual sparkplugs. With extensive vehicle operation, the coil can vibrate off and out of place, disconnecting it. The coil wire is one of the few components in a vehicle that can be taken out and replaced by hand.

Two other common repairs to the ignition system are **replacement of the starter module** and **replacement of the starter motor**. The **starter module** is a unit that is connected to the battery and to the starter motor. The starter module controls the voltage that goes into the starter motor. In some vehicles, the starter motor is prone to malfunction.

To replace the starter motor, the mechanic must first locate the starter motor's location within the maze of components and wires that surround the engine. Sometimes it is difficult to locate the starter motor because manufacturers often place the starter motor towards the bottom of the engine. Sometimes the gears in the starter motor will lock up, resulting in the vehicle's main combustion motor failing to start. A crow bar is useful to physically strike the starter with a sharp, yet calculated, hit. If this technique does not free up the starter motor, then it must be immediately replaced. To do so, the mechanic must disconnect the battery first, before working near the starter motor. Two bolts hold the starter motor in place and a variable wrench or socket wrench is an essential tool to do so. The all-important screwdrivers—both Phillips-head and flat head—are some of the most useful tools to have on hand to disconnect the wires to the battery. Larger screwdrivers can be used to pry up the contacts from the battery if they have rusted.

The Electrical System

Electrical Systems and Components

The **battery** is the heart of the vehicle's electrical system; it is where the electrical system begins and ends. The start of the electrical system in an automobile is a positive and negative battery cable connection. The flow of electricity from the positive to negative terminals of the battery is the essence of the electrical system. Electricity must complete a closed loop to the battery in order for there to be an electrical flow. Two cables are inherent to the function of the battery—one cable from and one cable to the battery. A negative cable attaches from the battery to the body of the vehicle. The greatest draw on the battery occurs when the ignition system starts.

Another essential part of the electrical system is a unit known as the **alternator**. The alternator is driven by the crankshaft. The spinning of the alternator by the crankshaft produces electricity for the vehicle while the engine is running. The vehicle computer monitors the voltage regulator. The starter's electric motor, which is connected to the battery, engages the combustion motor by starting the combustion process in the piston chambers. The starter electric motor turns the crankshaft, which then moves the pistons and simultaneously the sparkplugs spark, igniting the gas in the piston chambers. A large wire coming from battery to the solenoid is one part of the circuit and a thin wire attached to the vehicle body completes the circuit. The starter electrically grounds to the engine block via a wire.

Wires that comprise the vehicle's electrical system are fastened in many different ways, which is why connection failures can be a common malady with the electrical system. The connection points in a vehicle's electrical systems are the main culprits when problems arise. There are mechanical connectors on the battery that are comprised of clamps and nuts and bolts. In the electrical system, there are also plastic connectors and wiring harnesses that have many different wires within them, like branches on a tree. When the key is turned (or a button pressed), the electrical connection/ignition starts a current that runs from the battery to the starter, which then turns the engine's crankshaft. When the main combustion engine turns on and is operating, the engine powers a pulley with a rubber belt, which then powers the alternator, which charges the battery for the next ignition. The alternator produces electric current that powers the vehicle's accessories. The alternator generates electricity only because it is being powered or turned by the combustion engine.

If the alternator fails, then it must be replaced. A **tensioner** device is used to keep tension on the pulley bands. The rubber bands use a pulley system, so it is important for the mechanic to use a pry bar to get tension on the rubber bands that go over the pulleys.

The Drive System
Drive Systems and Components
While the engine is running, the flywheel turns the crankshaft. This moves the pistons in their respective cylinders. Any components located in the lower end the engine near the block, or components located inside the engine or located just below the engine block, are difficult to repair and require many tools. The use of an auto shop with a vehicle lift and full array of specialized tools is required to effectuate repairs to the lower part of the engine. The car's transmission transmits power produced by the engine to the wheels. The transmission is a complex component that requires special training to carry out needed repairs.

The Lubrication System
Lubrication Systems and Components
The oil contained in an oil pan below the engine is the "life blood" vat of the vehicle's lubrication system. When the pistons move, and the combustion engine starts operating, it starts the mechanical operation of the oil pump, which then circulates oil from the pan located below the engine. The oil moves up to the valves above. The oil coats the pistons then drops back into the oil pan. Lubrication prevents heat damage to the engine due to friction. The oil pump sits inside the oil pan. To replace the oil pump, the mechanic must drop the oil pan down by using different sized wrenches, and then the new oil pump can be reconnected.

Rocker arms are an essential part of the engine near the valves. Rocker arms require a steady, constant dousing of oil to work efficiently. A full pan of fresh oil is essential because old oil can gum up the works of an engine. The use of various wrenches facilitates the removal of the oil pan.

Cooling System
Cooling Systems and Components
An automobile engine commonly runs at an average temperature of 180 degrees Fahrenheit, which is the optimal temperature for the sound operation of most auto engines. The temperature of the engine's interior cylinder walls is 350 degrees Fahrenheit and the top of the cylinder is 600 degrees. The actual combustion fire in the cylinder combustion chamber is 3,000 degrees Fahrenheit.

In order to cool the engine, water circulates in channels throughout the engine. As the engine runs, the engine fan takes in air through the radiator. As the air passes by the fins of the radiator, the flowing water cooled by the air cools the engine. Cooled water flows through the engine jacket.

Hoses from the radiator carry water into the engine channels. Sometimes the hoses break or develop holes and must be replaced. Hose clamps connect the hoses to the engine and to the radiator. The hose clamps can be disconnected by needle nose pliers used on the hose clamp. To drain the coolant from the engine, a wrench is needed to unscrew the cylinder block drain plug. Some cars have bleed screws that must be opened using Allen wrenches with hexagonal nuts.

Engine temperature is sometimes controlled with the use of a unit called a **thermostat**. If the thermostat gets stuck open, the vehicle will overheat. If the thermostat gets stuck closed, it will not generate any heat for the car to use to heat the passenger compartment on cold days.

Cylinder head gasket problems require several tools, including a torque wrench. **Cylinder heads** must be torqued down to a specific pressure. Next, the intake manifold must be taken off and then the head fuel injection must be removed. All of the bolts must be removed and then the cylinder head can be lifted, and the old gasket should be scraped off. Head gaskets are made from a composite material. The head gaskets on a car's engine can be blown if the car overheats.

The Exhaust System
Exhaust Systems and Components
The **exhaust manifold**, also known as a **front pipe**, is designed to take gases and gas byproducts from the engine cylinders and channel them into the exhaust. Then the exhaust gas travels down an exhaust pipe to a catalytic converter, which is a filter designed to reduce harmful pollutant gases. The center section, also known as a **muffler** or **silencer**, is designed to reduce noise. The tailpipe further reduces noise and funnels exhaust gases away from the vehicle.

Exhaust gases produced by combustion in the piston chambers are channeled to the exhaust manifold. Without an exhaust manifold, a vehicle will run very loud and the engine will not function properly because there will not be enough pressure to keep the pistons operating. The exhaust manifold often gets worn out from the hot gases and damaged by vibrations emanating from the engine. Wrenches are necessary to replace the exhaust manifold. There is a gasket for the exhaust manifold known as the **exhaust manifold gasket**. After the exhaust manifold, there are clamps in the front of and to the back of the catalytic converter. The **catalytic converter** filters harmful gases from going into the environment. There is the muffler for the exhaust system that is also held in place by U clamps. A wrench is needed to take the muffler off the U clamps and each has two bolts. On some cars, straps hang the muffler and let the muffler sway with some give. Metal mounts also support most mufflers and extend down from the body, also giving the muffler system some sway and give. The give is advantageous because when the muffler gets hot; it expands and can split in places if totally bolted down. Other vehicles have a welded muffler system, which benefits from the steadfastness of a metallic weld.

Important tools to fix the exhaust system include ratchet sets, open end wrenches, and channel lock pliers. Channel lock pliers are adjustable, so if the mechanic encounters many different sizes of nuts to remove, the width between the jaws of the pliers can be adjusted. The jaws are typically an inch to two inches in length and extend at a right-angle from the two handles. The gap between the jaws is variable, due to a tongue-and-groove sizing system that lets the user adjust the jaw gap. Some channel lock pliers have rubber handles that protect the mechanic from electrical shock when working near wires that contain a live electric current and can be used to remove rubber supports that might be along the exhaust system. Channel wrenches and silicon spray can be used to help remove parts. Two bolts that hold the muffler to the tailpipe might be spring-loaded.

The Brake System
Brake Systems and Components
When the brake is depressed, the brake fluid flows through the master cylinder and then to each wheel's break cylinder. The resulting fluid pressure causes physical pressure, which creates friction. A hydraulic brake system is an extension of Pascal's Law, which states: "Pressure exerted anywhere in a contained incompressible fluid is distributed equally in all direction throughout the fluid."

The most common brake systems in a vehicle are **disc brake systems** and **drum brake systems**. The main components of a disc brake system are the wheel hub assembly, the brake caliper assembly, the disc brake rotor, the wheel, and the lug nuts. The wheel hub assembly holds the wheel and the disc rotor. The bearing inside the wheel hub assembly allows the smooth rotation of the wheel and the disc rotor. The disc rotor is the component that the brake pads squeeze against. This creates friction, which slows the rotation of the wheel. The disc rotor produces a lot of heat due to friction and that is one of the main reasons why brake pads wear out and need to be replaced.

The brake caliper assembly uses the hydraulic force from the brake pedal to squeeze the brake pads to the rotor's surface, creating friction and decelerating the wheel. The caliper assembly consists of the caliper bracket, the inner brake pad, the slider pins, the dust boots and rubber bellows, the outer brake pad, and the brake piston. The caliper frame has fittings that hold the hose that contains the fluid that goes to the piston within the caliper frame. When the brake pedal is depressed by the vehicle operator, the fluid flows from the master cylinder to the caliper piston. The fluid pushes the piston, which makes the inner brake pad squeeze against the rotor surface. The piston also exerts force on the caliper frame and outward force is exerted on the caliper slide pins, which enables the outer brake pad to squeeze the rotor at the same time as the inner brake bad. The caliper frame slides along the slider pins.

The **disc brake calibers** contain two brake pads. Disc brake pads can get worn by repeated braking of the vehicle. A jack to lift the car is required if a mechanic's garage lift is not available. Wrenches are needed to replace brake pads.

The lack of brake fluid can cause a vehicle's breaks to fail, so it is important to check the brake fluid reservoir in the car. Holes can form in the break pipes and hydraulic system leakage can occur. Brake fluid pipes must be replaced if there is any leakage of brake fluid.

Drum brakes are also used in many vehicles. The drum brakes are often used on the rear wheels of a vehicle. The important components of a drum brake system include:

- **Drum**: a round cast iron metal housing that looks like a bowl

- **Wheel cylinder**: the component that forces the brake shoes out towards the inside walls of the drum

- **Brake shoes**: these components press against the inside of the drum rim

- **Hub**: supports drum brake and wheel

- **Springs**: pull the brake shoes back away from the drum when the brake is released

- **Adjuster**: a small cylindrical part that regulates the distance of the brake shoes from the surface of the inside of the drum. As the shoes get worn, the distance between the shoe and the drum becomes greater and the drum break adjuster component keeps the distance constant. The adjuster has a small wheel that adjusts the distance when the brake is repaired and the shoe distance is set.

- **Strut**: keeps the shoes from compressing too far and keeps the shoes in alignment.

- **Backing plate**: holds the drum in place and also supports the drum brake cylinder.

Drum brakes usually have a short shoe and a long shoe. The short shoe goes in front, towards the front of the vehicle. The long shoe is placed more to the rear of the vehicle.

There are a few specialized tools that need to be used when fixing drum brakes: a multi-spring tool (which looks like large forceps), an adjuster tool, spring removal tools that have special ends that help remove the springs (these tools are important because the springs have a lot of force and could injure the mechanic if they pop off), a screwdriver, and needle nose pliers.

Before servicing the brakes, the mechanic must park the vehicle on a level surface and make sure that the parking brake is not engaged because this will press the brake shoes against the drum and prevent the brake drum from being removed. The mechanic must always place cars with automatic transmissions in park and those with standard transmissions in neutral before working on the rear drum brakes. The first step is to pop open the hood of the vehicle and check the brake fluid reservoir to make sure that there is the proper amount of brake fluid, which should be filled to the "full" line. This is the fluid that goes to the master cylinder and it should appear clear without debris. If the fluid is cloudy or dark, it is recommended that it is replaced. Before taking off the wheel, the lug nuts on the wheel must first be loosened and only then should the vehicle be jacked up. If a lug nut is rusted on or too tight to take off without using great force, then the vehicle should not be jacked up. If great force is applied while the vehicle is jacked up, the vehicle can slip off of its supports. The drum brake contains many parts and they can be difficult to reassemble if there is no reference to their proper positions. The mechanic should take a picture of the drum brake before any disassembly. A rubber mallet can be an essential tool if the drum is stuck and will not come off. A few taps with the rubber mallet should loosen up the drum for removal. Some drums use retaining screws, and a screwdriver must be used before the drums are removed. A brake adjusting tool is essential. The brake adjusting tool resembles a small, hand-size crow bar. The brake adjusting screw is an important component of the drum brake and must be loosened before attempting to remove the brake shoes. A brake spring washer tool is a good addition to the mechanic's arsenal for removing the springs in a drum brake. When replacing the wheel after working on the brake drum, a ratchet and socket tool is beneficial for ensuring that the lug nuts on the wheel are tight, after which a torque wrench is used to achieve optimal lug nut tightness.

Suspension System
Suspension Systems and Components
The **suspension system** of a vehicle helps provide a smooth ride when driving. Vibration can cause a host of maladies within a car and can literally shake a car to pieces if uncontrolled. The less vibration that a vehicle experiences, the fewer the necessary repairs. The other important factors associated with the suspension system is that the system keeps the vehicle upright and supports the vehicle from excessively tilting and swaying when the vehicle is turning, accelerating, and braking.

Suspension systems are comprised of many components. The main components are springs, shock absorbers, and control arms. The control arms are attached to the vehicle's chassis frame. The springs most used in vehicles are coiled springs because their shape permits them to be mounted directly over the shock absorbers. Some vehicles use leaf springs (flattened bands of metal that are stacked) for the rear wheels of the vehicle.

Shock absorbers are also referred to as dampers. Modern shock absorbers look like tubes that are usually gas-filled (older shock absorbers were oil-filled) within the tube and dampen the vibrations of the road. The shock absorber also dampens the oscillations of the springs and offers a positive and negative resistance, in both directions up and down to the springs, yielding a more stable ride.

The **front suspension** components consist of arms that are frequently referred to as wishbones, hence, the term **double wishbone suspension**. The arms of the wishbone usually have one to two ball joints that enable steering and vertical movement. The lower arm of the wishbone supports the spring and shock absorber, while the upper wishbone supports the steering component to the wheel. A comparable suspension system is known as the **MacPherson strut suspension**. The MacPherson system is desirable and offers some advantages. In the MacPherson system, the upper wishbone arm is replaced by a strut (a longer shock absorber body), which enables grouped functions in one unit; the strut also acts as the upright support for the spring and alleviates the need for another wishbone arm. This simpler design adds to passenger comfort.

An **anti-roll bar** attached between the shock absorber components of both front wheels resists the rolling that occurs when turning. The inertia of the turn places force on the outer springs. The anti-roll bar presents a counterforce that maintains stability against the vehicle's sway.

Rear suspension systems in vehicles usually have an independent configuration where each rear wheel can move up and down independently from the other rear wheel. A non-independent rear suspension system has a solid axle connecting both rear wheels. This is not as desirable as an independent configuration because if one wheel hits a bump, the other wheel is also affected, because the sole axle transmits the force of the bump to the other wheel. The key component of the independent rear suspension system is the universal joint. Instead of one solid rear axle, each of the driveshafts extends from the differential (the unit that transmits power from the drive shaft to the axle) into universal joints. The universal joints enable the wheel axles to move up and down while the differential gear casing stays in place.

The Steering System
The Steering System and Components
All vehicles have a **steering wheel** that turns the front wheels of the vehicle. The front end of a vehicle usually has a lot of weight because the engine is located in front. In early vehicles, before the advent of power steering, it was sometimes difficult for the driver to turn the front wheels. With the help of the engine's power, steering has become almost effortless and very responsive. The steering mechanisms are usually powered by an **engine-driven pump** and a **hydraulic system**. With an older rack and pinion steering system, the rotational movement of the steering wheel is transferred to the frame mounted rack and pinion steering assembly. The **rack** and **pinion mechanism** offer precise steering abilities but require more effort to move the wheels.

The main steering components consist of a steering wheel, steering shaft, a steering rack or steering box, and arm mechanism. Universal joints are placed throughout the steering shafts so that the steering mechanisms can get around oddly placed components in the vehicle that are attributed to other vehicle systems. Also, the universal joints allow for repositioning of the wheel for the operator's comfort.

The **steering box component** is a gear box designed to make steering easier for the vehicle operator. These units often become loose and consequently result in a lot of unwanted "play" in the steering of the vehicle. The steering will feel loose and the operator may have to compensate by physically steering with increased revolutions. A new replacement steering box is easily installed but the mechanic must have a "stubby" short ratchet in order to access the congested part of the engine compartment. A short-handled ratchet is a tool that can fit where larger tools would be encumbered by limited space. A pry bar is a necessary tool to work surfaces apart. While mostly 34mm and 24mm nut sizes are encountered in the steering systems, mechanics should have an array of different sized ratchets and wrenches to adjust a variety of nuts and bolt sizes.

Automotive Components

There are several essential components of a vehicle that require replacement including:

- **Brake Pads**: One of the most frequently replaced components in a vehicle are the brake pads. The use of a mechanic shop car lift makes the brake pad replacement process much easier because the mechanic can stand and work under the vehicle in relative comfort. The first step in replacing the brake pads is to take out the slide pins that the caliper rides on using a socket and wrench to remove nuts that hold the slide pins in place. Once the caliper is removed, the mechanic can keep the caliper away from the brake disc by letting the caliber hang with the use of a piece of wire hanger. Finally, the brake pads that are now visible to the mechanic should easily come off of the caliper. The replacement of brake pads is easy because they fit right over the disc and are supported by small appendages around the disc. The mechanic fits the caliper (the device that produces the pressure around the disc) over the brake pads and proceeds to perform the initial tightening of the caliper around the brake pads and onto the caliper support. The most important step is at the end of the brake pad replacement process: the tightening of the brake caliper using a torque wrench to exact specifications set forth by the vehicle's manufacturer. A torque wrench shows the exact amount of tightening by using a mechanical meter that displays foot pounds. For example, if Ford specified 30-foot pounds of torque was required for one of their car's brake calipers, then it would be most important to adhere to that specification. If the bolt is too tight, the caliper pin may break, or the entire brake system could become damaged. If the torque is insufficient, then the caliper pin might fall out. The mechanic must make certain that all hydraulic brake lines are back in place and then insert the spring clips that go inside the caliper. The spring clips lessen the amount of brake squeal by holding the brake pads in place using tension. The spring clips go inside the caliper and the clips are specific to the vehicle manufacturer.

- **Drum brakes**: The shoes of the drum brake often need replacing and a few specialized tools must be employed, such as an "adjuster" tool and a "spring" tool. The complete description of replacing drum brakes is a complex process and beyond the scope of the exam.

- **Valve cover gaskets** keep gas and oil from escaping onto the surface of the engine and causing a smoking condition. If replacement is necessary, the valve covers must be unbolted and new gaskets placed along the rim of the valve cover base.

- **Tires**: The wear on a tire depends on the mileage driven on the tire and the type of tire on the car. A softer tire will adhere to the road better but will have a much shorter life than a hard tire with a deep tread. Tires are usually replaced after 25,000 to 60,000 miles of travel.

- **Lights**: Head lights, tail lights, indicator lights, and interior lights of a car must be changed when burnt out or damaged

- **Hoses**: Radiator hoses, transmission hoses, and brake lines must occasionally be replaced. Just three tools are necessary for fixing a broken hose: a flat head screwdriver, a cutter type scissor tool, and channel lock pliers. After the screw holding the hose in place at the source is removed by using the flat head screwdriver, the mechanic has an opportunity to further inspect the hose. If the hose damage permits cutting the damage away and using the remaining length of hose to reconnect the hose, then a new hose might not be required. A complete hose replacement requires loosening the hose band on the other end of the hose and removing the hose. A hose

band is a metal ring that goes around the hose and secures the hose to the mouth of where the hose must be placed. The hose ring is tightened with either a slotted screw or small nut.

- **Wires**: Heat sometimes melts the wires in a vehicle. A short in the electrical system of a car can lead to a dangerous fire condition. Wires must be inspected and if worn, must be replaced.

- **Electrical connectors**: Age and heat can rot out an electrical connector. Simple tools are used to replace these components.

Procedures for Automotive Troubleshooting and Repair

The first step to auto troubleshooting is to be aware that there is a problem with the functioning of the vehicle. Sometimes a vehicular problem can be detected just by listening to a sound that is discordant, such as pinging, the grating sound of a worn brake pad against the rotor of the wheel, or a knocking sound when the vehicle's wheel is turned. This latter sound can indicate that the ball joints are loose in the steering mechanism. Another way to ascertain if there is a potential vehicular problem is to keep acute attention on the gauges of the car. If the temperature gauge on the vehicle's dashboard is reading close to a red-hot zone, it can indicate that the vehicle is lacking coolant or perhaps that the pump for the coolant is not operating properly, causing the engine to become hotter than normal.

Puddles of Liquid Under the Vehicle

The vehicle operator and vehicle mechanic must be cognizant of any puddles of liquid that have formed under the vehicle. These puddles are caused by leaks in vehicular systems. The motorist should take note of these puddles as they might spell significant vehicle operation trouble down the road. The type of liquid that has accumulated must be determined first. For example, it may be innocuous rain or condensation from the vehicle's air conditioning system.

Transmission fluid is a purplish color and the sight of this puddle can indicate a serious problem. The mechanic should touch the fluid with the index finger and thumb. If the fluid is slightly viscous and purple, then it is most likely transmission fluid. An immediate inspection is necessary to determine the source of the transmission leak. The transmission is a very expensive unit and to avoid replacement, the motorist must repair the leak. If a vehicle's transmission fluid becomes too low, then gears within the transmission can grind together and ruin the transmission to the point where it must be replaced.

Leaking brake fluid also necessitates immediate repair. If the brake fluid is too low in the brake fluid reservoir, then the vehicle will not stop, which puts the motorist and others in grave danger. **Brake fluid** is slightly viscous and darker in color. The mechanic must determine from where the brake fluid is leaking and then plug the leak. The master cylinder or one of the many tubes that bring brake fluid to the wheels might be leaking.

Antifreeze is greenish-yellow and has a distinct smell. When a coolant pipe leaks or a vehicle's radiator has sustained a small hole and is leaking antifreeze, the coolant liquid contacts the hot engine. The coolant liquid, commonly known as antifreeze, will smoke when it comes in contact with the hot engine. If an antifreeze leak is suspected, the coolant vat should be checked and if low, there may be a leak. The location of the leak can sometimes be detected by the location of the coolant puddle under the vehicle. Usually the location of the puddle is directly below the leak. From there, the mechanic can use a bright flashlight to locate the exact position of the leak. Various methods of repair can be used to remedy the coolant leak.

A leaking **oil pan** is usually detected when a car moves from a parking spot and there is a significant spot of oil where the car was parked. Specifically, there is an amount of oil detected on the pavement under the car's engine when parked. Any amount of oil detected on the pavement can denote a dangerous and costly condition that must be repaired. Even a small leak over time will deplete the oil pan of oil and consequently leave the engine without essential lubricant. The damage done to the engine can be catastrophic, resulting in an engine seizure. The engine is usually the most expensive part of a car and its loss is costly. Insurance companies will not cover the engine replacement in the case of an oil leak because the damage was perpetrated by the negligence of the car owner. When an oil pan leaks, it has usually been struck by a rock or some other protruding object that the underside of the vehicle has come in contact with.

If there is a hole in the oil pan, then it must be replaced. Usually just two or more bolts hold the oil pan in place and these can be loosened with the help of a ratchet tool. When replacing the oil pan component, the mechanic must make sure that the proper gaskets are available. All vehicle systems that use liquids usually have gaskets to prevent the liquid from squirting out through the connections within the systems.

Water condensation from the vehicle's exhaust tailpipes is not a concern. Most vehicles have water drips from the tailpipes when the vehicle is first turned on and operated for the first few minutes. The resulting natural accumulation of water condensation will remain in the tailpipe and cause rust to the entire outer half of the exhaust system. This only becomes problematic if the vehicle is only operated for a few minutes or less. Vehicles that often travel just a few miles at a time will experience a rotting out of the muffler and tailpipes because the condensation does not have ample time to dry out as it does with the extended use of the vehicle. To alleviate the possibility of ruining the exhaust system due to water condensation, a vehicle should be driven more than just a few miles to dry out the watery condensation that has formed in the exhaust system.

The Vehicle Lift

Proper operational knowledge of a car "lift" function is essential for the vehicle mechanic. A **car lift** enables the mechanic to operate under the car without having to lay supine underneath the vehicle. Moreover, the lift permits the removal and installation of larger components under a vehicle. Working under a four-thousand-pound car can be extremely dangerous if the mechanic is not familiar with the safe operation of the car lift. Car lifts usually have adjustable arms that have rubber pads, which rest against the underside of the vehicle at the frame's strongest points. Care must be taken to place the rubber ends of the lift's arms in the proper position at the strongest points of the underside of the car where the weight of the car can be supported without damaging the vehicle. The car is raised to a point where the mechanic feels that he or she can comfortably and safely work under the car and then the lift is locked in place.

An adequately equipped machine shop or vehicle repair shop contains many tools including the following:

- A **Metal Inert Gas** (MIG) welder is useful for bonding metals. MIG welding is a form of electrical welding that requires a safe workspace away from ignitable fuels. The shop welder must wear a welding helmet to protect against hot flying sparks and flames and an air ventilation and exhaust extraction system is necessary when welding indoors. For safety, the MIG welding machine should be of quality design.

- A hand-operated **Sawzall™** (a name that is a trademark of the Milwaukee Electric Tool Company) is an important shop tool. This automatic electric hacksaw or jigsaw can cut many materials, including metal. It is essential for removing unwanted parts and for fabricating parts. The newer Sawzall™ operates by long lasting batteries that remove the need for cumbersome electrical cords.

- A **stout hand drill** with a lot of torque is essential for drilling through steel as well as concrete. A diversified drill set enables the mechanic to work on different types of repairs.

- Assorted **Cobalt wrenches** arranged neatly by size on a wall or in a tool box will facilitate the work of the mechanic. These are strong fixed wrenches that work on many sized nuts.

- **Wire strippers** are necessary for electrical work. An assortment of wires and electrical connectors should be available in the shop.

- A **rubber mallet** should be used when the surface of the object being struck would be damaged with a regular metal hammer.

- A box of **assorted steel nuts, bolts,** and **washers** organized in a tool box.

- A **carbide blade chop saw** enables the shop mechanic to cut metal and wood. A chop saw is a circular bladed saw that is on a hinge. Items are placed under the blade and the operator lowers the circular saw to cut the object.

- A **scrap bin** should be used for disposing of metal shavings and other unwanted material refuse produced from working with materials. Sharp metal pieces lying around the shop floor can cause injury, so they must be removed. A scrap bin made of a heavy gauge metal or plastic is necessary to contain the refuse produced by a machine shop.

- A **multi-drawer tool box** can improve efficiency; time is always an important factor when considering a repair or part fabrication. The tool box enables the mechanic to have instant access to the necessary tools in an organized fashion. Sometimes, tools can become misplaced and unnecessary time might be spent looking for the proper tool. The tool box lets the mechanic sort out the various chisels, sockets, hex keys, wrenches, crescent wrenches, various pliers, and other multiple tools. A well-sorted multi-level, multi-drawer standing tool box enables the mechanic to find the exact tool required and act with it to complete the job at hand. Drawers with smooth ball bearing action let the mechanic open and close the drawer quickly and easily.

- **Metal files** are often important for trimming components. Various grades of roughness give the mechanic options when shaving materials to specific sizes.

- The value of **simple shelving** in the shop space should not be underestimated. Shelves permit the mechanic to see where items are located, improve accessibility to such items, and can support heavy cans of paint and other tools. Shelves help prevent damage to expensive equipment by getting it off the floor or out of messy piles. Metal shelving can easily be constructed and secured against the walls of the mechanic's shop.

- **Dial calipers** make important precise measurements. They are easy to use and are essential for measuring the width of materials to be worked on.

- An **angle grinder** (side grinder or disc grinder) lets the mechanic shape a variety of materials depending on what type of circular brush is used. It can be used for cutting, grinding, and polishing. Some angle grinders are on a stand while other angle grinders are handheld. They are powered by an electric motor, gas engine, or compressed air.

- A **filtration system** to expel bad air is essential for the shop. Fume extraction and oil smoke extraction as well as dust and oil mist must be sucked out of the air by a fume extraction system. These units are essentially big standing vacuums that pull the toxic air away from the mechanic.

- A **Tungsten Inert Gas** (TIG) welder uses a tungsten electrode that delivers the current to the welding arc. The MIG process is easier because it uses a continuously feeding wire whereas TIG uses long welding rods that slowly feed into the weld puddle. TIG welding is used on metals of thinner gauges.

- **Mill drills** are used very frequently in all types of machine shops. A mill drill is a large, electrically-powered vertical drill that drills down vertically with a bit. Milling is a machining process that uses rotary cutters to remove material by feeding in a direction at an angle to the axis tool. Milling is one of the most commonly used processes for machining parts to precise sizes and shapes. The shop's scrap bin should be accessible as this machining process produces a large amount of metal waste.

- A **lathe** is an essential tool that will conform metals to size and shave surfaces to the desired dimensions. It is similar to a mill drill but is based on working with a horizontal axis. The lathe is a machine for shaping material by means of a rotating drive that turns the material piece being worked on against changeable cutting tools. A lathe can be used in the manufacturing of various machine components.

- An **air compressor** is a useful unit to have in a machine shop because some tools operate off of compressed air, such as wheel nut wrenches, which are used to remove tires from the wheel of a car. Air compressors are also useful for inflating tires and blowing out debris and dirt from machinery and small spaces.

Construction Procedures

Concrete is an advancement that has revolutionized the ability to build comfortable residential structures using materials that are fantastically resilient to the elements. The structures built with brick, stone, and concrete can outlive the length of a human life. There are brick structures from 500 years ago that are still standing today, and some stone and concrete built houses can last for thousands of years.

When building a brick or stone building, the concrete mix must be of a perfect consistency, much like peanut butter, and not too loose. A veteran bricklayer will spread the concrete evenly on top of a brick like spreading peanut butter completely over a piece of bread. Bricks are an important exterior material used in different applications all over the world, but the red brick is the predominant type of brick in the United States. When constructing a brick wall, the bricks must be straight and level, so that the force of gravity does not disturb the wall over time.

Constructing Brick Walls

To **construct a brick wall**, the mortar is spread completely over the top of each brick with a trowel or spatula tool. These flat tools are used to smooth out the mortar used between the bricks and to wipe away excess mortar that spills over the sides of the brick. The brick wall should be free of mortar on the sides of the bricks. One layer of bricks should be laid down and each brick should be tapped with the end of the trowel's handle. A "spoon" tool is used to create a concave indentation in the concrete that surrounds each brick. A bubble leveler is a very important tool and should be used consistently through the building of a brick wall.

Balloon frame or "stick built" houses require the use of a circular saw and many two-inch by six-inch boards. When building a house, only 2"x 6" boards are allowed because this size allows enough space for insulation to be installed.

The Roof

A waterproof, weatherproof **roof** on top of a structure is an essential part of the structure that must be resilient. There are just a few important tools and techniques to employ for the construction of a new roof. A hammer, a bucket of nails, a sufficient number of shingles, and tar paper are the main materials that are required. A flat shovel or the front of the spade can be used to get underneath the old shingles of the roof to shovel them off and expose the base roof plywood. Then tar paper is put over the plywood using a staple gun and then shingles are added from the bottom of the roof to the top. Roofing nails with big heads and shafts long enough to go through the tar paper, shingles, and plywood must be used.

Foundation

A poured concrete **foundation** is standard for the first stage in building a structure in locations where temperatures can often go below freezing. **Rebar**—iron/metal rods that extend through the concrete—are used as a reinforcement for the foundation. **Wood frames** can serve as molds for cement walls and are easily built using wooden beams, a circular saw, a hammer, and nails. Cement does not stick to wood, so wood is a good foundation framing material. After the concrete hardens in the frame, then the wood easily comes off away from the concrete. Concrete wall thickness is determined by the size and design of the intended structure but 18" thickness is the norm. The foundation should be at least 46 inches below the ground line. In locations that have temperatures that plummet below freezing, the posts and foundations must be buried below the frost line. In Florida, it is legal to build a house on top of slab; houses can be built on slab anywhere where the temperature does not dip below freezing.

Below 17 degrees Fahrenheit, concrete should not be poured. When temperatures dip down, it is good to add antifreeze to the water. If water freezes, then the concrete will not cure. Tools used for constructing a foundation include a wheel barrow for transporting the cement to the immediate location where it is needed, a shovel to push the concrete into spaces, and a cement "squeegee" to level the cement once it is poured into place.

Portals: Door Frames and Window Frames

Typically, either 2" x 4" or 2" x 6" boards form a window sill. Four boards support the sill; these boards are also referred to as **studs**. Trimmer studs extend vertically from the bottom horizontal stud and these form the sides of the window frame and are nailed into the king stud, which is one of the studs that make up the main frame of the wall. The trimmer studs support the **header assembly**, which is the top of window frame. The header assembly is four different individual pieces of wood and often must conform to building code regulations within the structure's municipality. Two 2" x 6" boards are recommended for the riser part of the header. Between the top of the header and bottom of the upper

plate (top horizontal stud), fitted studs—usually three—brace the header against the upper plate. A row of two nails every sixteen inches should secure the trimmer studs to the king stud. If using 2" x 6" framing, then three nails should be used every sixteen inches along the sill support studs next to the trimmer studs. Nails can be used to secure the horizontal sill to the vertical supporting studs. A drilled hole with a bolt can also be installed horizontally through the king stud and the trimmer stud into the vertical sill. The header should have four nails through the king stud into the riser part of the top window frame.

A **door frame** is very similar except that there is no triple support from below because the door goes all the way down to the floor. The trimmer studs on either side of the portal make up the door frame sides. The trimmer studs support the top or header of door. The door header should have rows of three nails every sixteen inches. The header of a door should be strong because it needs to support anything that is above it. It is recommended that a precisely fitting sheet of plywood is used between the header riser boards (the boards that are placed facing forward). A board nailed onto the bottom of the header will lead to a nice flush, flat surface.

Practice Questions

1. What takes place during the upstroke of an engine's piston?
    a. Gas is cooled so the engine can achieve power.
    b. Coolant is injected so the engine doesn't overheat.
    c. Gas is compressed and exploded with a sparkplug.
    d. Oil is filtered around the exhaust.

2. What component generates electricity while the engine is on?
    a. Headlights
    b. Solenoid
    c. Coil
    d. Alternator

3. If a puddle of liquid is observed under a vehicle, what is the correct action?
    a. Mark the day on the calendar when the car should be serviced.
    b. Identify the liquid under the car and act accordingly.
    c. Taste the liquid. If the liquid tastes like oil, the oil pan needs replacing.
    d. Water from the air conditioner drains the coolant, indicating the radiator needs filling.

4. What component is the first in line from the engine in a vehicle's exhaust system?
    a. Muffler
    b. Exhaust Manifold
    c. Catalytic Converter
    d. Tailpipe

5. What are the most common types of brakes in a car?
    a. Disc brakes and drum brakes
    b. Air brakes and electric brakes
    c. Reverse brakes and forward brakes
    d. Induction brakes and engine brakes

6. Which suspension system component enables the rear wheels to move independently from each other?
    a. An axle
    b. A universal joint
    c. A damper
    d. A ball joint

7. Regarding a combustion engine, where is the oil pump usually located?
    a. In the oil pan
    b. Above the engine
    c. Next to the gas tank
    d. In the valve cover

8. Transmission fluid is typically what color?
   a. Green
   b. Purple
   c. Brown
   d. Black

9. Which vehicle component often needs to be replaced on a regular basis?
   a. Rack and pinions
   b. Brake pads
   c. Emergency brakes
   d. Window seals

10. A master cylinder is an important component in which vehicular system?
    a. Electrical
    b. Transmission
    c. Engine
    d. Brake

11. Which of the following liquid leaks needs to be stopped immediately?
    a. A clear liquid underneath the engine bay
    b. A purple liquid underneath the transmission
    c. A clear liquid coming out of the exhaust pipe
    d. None of the above

12. Which of the following is responsible for transferring power from the drive shaft to the axle?
    a. The differential
    b. The master cylinder
    c. The steering column
    d. The accelerator

13. How does the turning or pressing of a car's ignition key start the car?
    a. It provides the initial spark that the motor uses to create ignition.
    b. It opens up the fuel tank, allowing for the fuel delivery necessary for combustion.
    c. It completes the electrical connection needed to send the battery's electrical current to the ignition motor, which starts the motor.
    d. It turns on the battery, allowing for electric current to start the vehicle.

14. What are channel lock pliers?
    a. A fixed set of pliers that can tune the channel of a car's radio
    b. A funnel that channels oil into the engine of a car
    c. A type of adjustable pliers that have a right-angle positioned jaw
    d. A straight jaw pliers that are used to unlock car doors

15. When building a roof for a house, which procedure is the correct construction practice?
    a. Lay tar paper over the bare plywood
    b. Create space between the shingles so water flows into the side ducts
    c. Place bricks under each shingle for extra support
    d. Use short nails through the shingles to prevent damage

16. Which important part of a window frame must conform to building code?
    a. The sill
    b. The trimmer
    c. The stud
    d. The header

17. If a mechanic must cut off a car's exhaust tailpipe, what tool should the mechanic choose?
    a. A lathe
    b. A chop saw
    c. A Sawzall™
    d. An axe

18. What is the most important construction feature when building a brick wall?
    a. The bricks in the wall must be spaced evenly.
    b. The wall must have a deep four-foot foundation.
    c. The wall must be straight up and down and level.
    d. The wall must not have excess concrete.

19. What type of screw has a cross-slotted head?
    a. Allen
    b. Single slot
    c. Hexagonal
    d. Phillips

20. Which of the following is used as reinforcement in concrete?
    a. Lag bolts
    b. Rebar
    c. 2" × 4" boards
    d. Galvanized pipe

21. What is the purpose of a lathe?
    a. To grind down rough surfaces
    b. To drive long screws into wood
    c. To shave or shape a material such as wood or metal
    d. To bend metal to form the needed angle

22. Which of the following is necessary when welding?
    a. Air ventilation
    b. A hard hat
    c. Insulated clothing
    d. Safety glasses

23. How are shingles added to a roof?
    a. From left to right
    b. From right to left
    c. From bottom to top
    d. From top to bottom

24. Which of the following is used in brick laying?
    a. Cement
    b. Trommel
    c. Plaster
    d. Trowel

25. Which of the following is a feature of MIG welding?
    a. Use of a tungsten electrode
    b. Use of a continuously feeding wire
    c. Used on thinner gauge materials
    d. Used in an air-tight room

Answer Explanations

1. C: During the full cycle of an engine's piston, the piston's upstroke occurs as the piston travels up in the cylinder chamber, which decreases the space in the chamber, compressing the gas that has been injected in the chamber. When the piston has compressed the gas in the piston chamber, the gas will explode due to the hot electrical spark of the sparkplug that occurs during the apex of the upward motion of the cylinder head. The spark explodes the compressed gas, driving the piston downward from the force of the explosion in the piston chamber, which then mechanically transmits that power down to the transmission system of the vehicle. Choice *A* is incorrect because gas must be exploded by the spark of the sparkplug to achieve engine power. Cool gas does not produce power; only exploding gas produces power. Choice *B* is wrong because vehicle coolant has nothing to do with the piston. The power that the engine produces is derived from the explosion in the piston chamber. Coolant flows through channels in the engine block but the process has nothing to do with engine power. Choice *D* is also incorrect. If oil ever did leak onto an exhaust pipe, the oil would smoke, due to the extreme heat of the exhaust pipe caused by hot engine exhaust gases.

2. D: Electricity is created by the alternator's internal opposing magnets spinning past each other, creating a current. The alternator is turned by the power of the combustion engine and this also creates the motion. Choice *A* is incorrect because headlights use, rather than generate, power. Choice *B* is incorrect because the solenoid, which is in the starter system of the vehicle, controls electrical flow; it does not produce electricity. Choice *C* is wrong because the coil directs electrical flow; it does not produce electricity.

3. B: Identifying the type of liquid under a car can be the difference between driving away with no consequence or burning out the engine and causing an expensive repair. If the puddle is oil, then the engine may be in danger of burning out. If the puddle is just water from condensation produced by the vehicle's air conditioning system, then no action needs be taken because this water is normal. *A* is not the correct answer because the date marked on the calendar may be too late; damage might have already occurred. If the puddle is due to an oil leak, it can be an indication that the engine's oil pan has no oil, which can severely damage the engine. Some leaks, like an engine oil leak, must be addressed and repaired immediately before the vehicle is driven. Choice *C* is incorrect and dangerous. Contrary to what is shown in some movies, vehicle liquids should never be tasted. Vehicle liquids are toxic and tasting a liquid to identify it is not a sound option. Choice *D* is incorrect because the air conditioner does not drain the engine coolant system. Sometimes, water condenses due to the proper functioning of the vehicle's air conditioner, but this occurrence does not adversely affect any vehicle system.

4. B: The engine produces exhaust gases that must flow out from the engine. The exhaust manifold is the component that is placed next to the engine to capture the exhaust gases coming out of the engine. Due to the heat of the fresh exhaust exiting the engine, the exhaust manifold must be replaced every few years. The hot gases wear down the exhaust manifold component and engine vibrations erode the connection between the exhaust manifold and the engine. This causes the exhaust gases to escape directly from the engine, causing a very loud exhaust noise. *A* is incorrect because the muffler quiets the exhaust gases and is located more towards the rear of the vehicle, before the tailpipe. *C* is an incorrect choice because the catalytic converter lessens the engine's exhaust gases' effects on the environment, and is usually located somewhere midway under the vehicle, before the muffler. *D* is not correct because the tailpipe is the last component in the vehicle's exhaust system. The exhaust gases leave the vehicle through the tailpipe, which is located at the very rear of the vehicle.

5. A: Disc brakes and drum brakes are the most common type of brakes in a car. Most cars use the disc brake system on the front wheels. The rotor is a circular component of metal that is pressed by brake pads on both sides and this friction is what slows and stops a vehicle. Drum brakes also rely on friction; the drum brake shoes also slow and stop the vehicle using friction. *B* is incorrect because air brakes are mainly used on large trucks or buses. Electrical brakes are sometimes used on railroad systems. *C* is not the correct choice because reverse brakes are fictitious. Brakes always slow and stop forward and reverse motion of a vehicle. *D* is incorrect because induction brakes are sometimes used to stop certain electric motors. Engine braking is the act of using bigger, lower gears to slow the car down. This will slow down a car but does not actually utilize any form of braking systems.

6. B: The universal joint permits independent wheel suspension by splitting the axle, allowing the vehicle's occupants to experience a more stable, less bumpy ride. For example, if a right rear wheel travels over a big rock, the shock absorber will work independently just for that wheel. The left rear wheel will not be affected by the motion of the right wheel because the universal joint in the middle of the axle allows independent motion of the left and right rear wheels. Choice *A* is incorrect because an axle is a stiff, one-piece bar that the wheels are attached to, without any movable joints. Choice *C* is incorrect because a damper is a shock absorber that is positioned for each individual wheel and does not have anything to do with the wheels moving independently from one another. Choice *A* is not correct because a ball joint is integral within the vehicle's steering system and is not directly involved with the suspension system.

7. A: The oil pump is often located within the oil pan. The pump draws oil directly from the oil pan located below the engine. The placement of the oil pump within the oil pan enables the pump to easily pump oil directly to the engine located above. Choice *B* is incorrect because oil flows up from the bottom of the engine for a variety of reasons. Gravity is a primary reason; if the pump was located above the engine, then the oil would flow down from the top and then it would have to be pumped up to the top again, which would be a very inefficient design. When a car requires its oil changed to fresh oil, a mechanic simply loosens a nut on the oil pan underneath the engine and the oil flows out naturally. New oil is then poured in from the top of the engine. Choice *C* is wrong because gas tanks are usually located to the rear of the vehicle. It would be an illogical design to have the oil pumped to the front engine from the rear of the vehicle. Choice *D* is incorrect because the intake and exhaust valves over the pistons are located on top of the engine. The oil pan and pump are located towards the bottom part of the engine. This design also allows the mechanic to effectuate oil changes because the oil pan is easily accessible.

8. B: The color of transmission fluid is purple. If a leak deprives the transmission from an adequate amount of fluid, the entire transmission can burn out and lose all functionality. The transmission is typically the second-most expensive component in a vehicle to replace (after the engine). The Choice *A*, green, is incorrect. Green is often the color of anti-freeze, so if a green puddle is observed underneath the vehicle, it is most likely a coolant leak from the radiator, which is a serious problem. The engine can overheat and seize-up, totally disabling the vehicle. Any amount of leaking in the coolant system must be fixed immediately. Choice *C* in wrong. Transmission fluid can turn brown if it is old and needs to be replaced with new transmission fluid. Brake fluid can also turn brown if it is old and needs to be replaced. The color of transmission fluid should be regularly checked by using the transmission fluid dip stick. The color of transmission fluid should always be purple. If a brown color is observed, then the transmission fluid is old and needs to be changed. The number of miles between transmission fluid changes depends on the make, model, and year of manufacture of the vehicle. Choice *D* is incorrect because black fluid often denotes the color of oil. If a black puddle is observed underneath the vehicle, then service to the vehicle must be performed immediately, on the spot where the vehicle is parked. Severe damage to the engine can occur if the oil pan has an insufficient amount of oil.

9. B: Brake pads often wear out from normal driving and must be replaced. When brakes are applied, the pads contact the rotor, stopping the movement of the vehicle by employing friction on the brake disc of the wheel. The friction causes wear on the brake pad. Over time, the brake pads will totally wear down, causing the brake caliper to contact the brake disc, which leads to brake disc damage. Choice *A* is incorrect because rack and pinions are steering components that do not wear out quickly and have a useful life over a period of many years. Choice *C* is incorrect because the emergency brake is rarely used when the car is in motion. Most drivers use the emergency brake to lock the car into a stationary position, especially if the car is parked on a steep hill. The emergency break is intended to keep the vehicle from rolling while the vehicle is supposed to be in a stationary, parked position. Therefore, there is very little wear because the car is already stationary when the emergency break is employed and it is a system that usually locks the rear drum brake(s). Choice *D* is wrong because the rubber around the car's windows usually lasts for the duration of a vehicle's lifetime; they rarely need replacement.

10. D: The brake system is arguably the most important system in any vehicle. If a car cannot "go," it is usually not an immediate safety issue. However, if the brakes fail to stop a vehicle's motion, then physical damage and serious injury or death can result. The master cylinder holds brake fluid and pushes the fluid into the brake calipers when the brake pedal is depressed. If the brake fluid is not present in the master cylinder, then the brakes will not operate, as brakes depend on the principles of fluid dynamics to work. Choice *A* is incorrect because the electrical systems of a vehicle are not liquid-dependent systems. Choice *B* is also wrong. While transmissions use fluid dynamics to work, transmission systems do not have a component called the master cylinder. Choice *C* is incorrect because there is no component in the engine system that is called the master cylinder. The engine system of a car deals with gas and oil fluids; however, the term **master cylinder** is generally used only for a component attributed to the vehicle's brake system.

11. B: The purple liquid would be the transmission fluid. If this leak is not stopped and repaired, the entire transmission can be destroyed, causing costly repairs. Choice *A* would just be condensation from the air conditioner, which is a normal occurrence. Similarly, Choice *C* is also a regular occurrence, although the car should be driven more than a few miles so that the water is evaporated and the exhaust pipe does not rust.

12. A: The differential is the set of gears that transfers power from the drive shaft to the axle. Choice *B* is the cylinder that contains brake fluid. The steering column is a shaft that connects the steering wheel to the steering mechanism, so Choice *C* is incorrect. Choice *D*, the accelerator, is the far right pedal that controls the amount of gas that flows into the engine allowing for higher speeds.

13. C: It finishes the circuit that powers the ignition motor, or starter motor, which turns the crankshaft, which causes the cylinders to move up and down while valves above the cylinders inject fuel. Once the spark plugs ignite, the combustion process begins and the engine starts.

14. C: The jaws on these pliers are positioned at a horizontal right-angle, which enables the mechanic to grab onto a nut or bolt that would be hard to turn without the right-angle jaw feature of the channel lock pliers. The word *channel* in channel lock pliers denotes the tongue and groove adjustment feature on the pliers. Choice *A* is wrong because it describes pliers that are fixed. Choice *B* is incorrect because a funnel is not a type of pliers. Choice *D* describes a tool that has straight jaws but those on the channel lock pliers are at a right-angle for extra force and maneuverability.

15. A: Laying tar paper over the plywood on a roof protects the plywood, which would otherwise absorb water. The tar paper is essentially another layer of waterproofing under the shingles. After the tar paper is covering the surface of the plywood completely and stapled to it, the shingles are nailed to the tar paper. *B* is not correct because there should never be any space between the shingles because this would allow water to permeate between the shingles. Rain water and melting snow should flow over the shingles and then flow into gutters built around the perimeter of the roof. The gutters are designed to direct the water away from the foundation of a house. *C* is incorrect because the use of bricks for a roof is not a correct construction practice because bricks are too heavy for a roof and can collapse the roof into the house structure. *D* is incorrect because short nails are not of sufficient length to go through the shingle, the tar paper, and finally, the plywood of a roof. Only long nails should be used for securing shingles. Long nails hold the shingles securely to the tar paper and plywood.

16. D: The top of the window frame, the header, often must bear weight and is under the municipal umbrella of construction rules. This top part of the window frame is also the most complex, consisting of the most wooden boards or other material used. Choice *A* is incorrect because the sill, the bottom part of the window, while important, does not have to conform to municipal code building standards. *B* is not the correct choice because the trimmer extends down to the base, usually the floor, and due to its vertical nature, does not fall under a code statute. Also, the trimmer is a very strong, stable upright vertical component of the window frame as a result of its location and attachment next to the stud. Choice *C* is not correct because the stud is a board that extends vertically in the wall and is not part of the window's frame.

17. C: The battery powered Sawzall™ is a tool that can cut metal easily and also get into tight spaces. It is hand-operated and it gives the mechanic the flexibility to reach into tight spaces in the vehicle. Choice *A* is wrong because a lathe is a machine that skims and trims surfaces to desired dimensions. It is a large machine that is not suitable for cutting tasks while positioned under vehicles. Choice *B* is incorrect for several reasons. A chop saw is a circular saw with a hinged arm, used mainly for cutting wood with its large "teeth." Therefore, a chop saw blade is not suitable for cutting metal. Also, it is a large, heavy tool with a metal base that is very difficult to maneuver under a car. Choice *D* is incorrect because, while an axe might ultimately work in an emergency, it is not suitable for cutting off a metal tailpipe. The axe would mangle the steel and not produce a clean cut. Moreover, a stout striking surface must be placed behind the tailpipe when striking the tailpipe with the axe to protect other components behind the tailpipe from damage.

18. C: The most important factor in the construction of a brick wall is ensuring that the wall is straight and level. Gravity plays a crucial role in the construction process and the longevity and soundness of a wall. The force of gravity falls in vertical lines and can strengthen a wall that's built straight. The slightest tilt of the wall from vertical enables gravity to exert forces that can exacerbate the angle and bring the tilted wall down over time. Choice *A* is incorrect because brick spacing is not as important as straightness in the structural integrity of walls. Although even brick spacing is a proper construction procedure, a brick wall will stand satisfactorily if the bricks are not spaced evenly because the concrete in between the bricks will continue to support and offer load-bearing cohesion to the wall. Choice *B* is incorrect because a wall does not necessarily have to have a four-foot-deep foundation to be able to stand soundly. A foundation for a brick wall is a good building practice; however, if the wall is just a few feet tall, then the brick wall should stand satisfactorily with a foundation that is less than four feet. Choice *D* is incorrect because excess concrete, while not aesthetically pleasing, will not affect a wall adversely. The brick wall will remain standing even though the concrete was sloppily applied.

19. D: The Phillips-head screwdriver has a cross-cut head, in order to turn the top of the Phillips-head screw, which has cross-cut slots on the head. Choice *A* is incorrect because an Allen head base is usually an indentation with a hexagonal or octagonal depression on the head of the screw. Special Allen wrenches are required for these screws. Choice *B* is wrong because the single slot screw head is just a single line depression on the head of a screw and not a double depression. The single slot screw is the most recognized screw and although useful, the cross-cut Phillips-head screw is easier to use with a drill because the Phillips screw holds the bit of the Phillips screwdriver more securely, due to the increased surface area. Choice *C* is incorrect because hexagonal heads are similar to Allen screws; however, they often have a star-type pattern indented on the screw head.

20. B: Rebar is a long metal rod. These rods are used in foundations to provide reinforcement to the concrete. Choice *C* is wrong because concrete would not adhere to wood, so that wouldn't help. Also, Choice *A* is wrong because lag bolts can be used to secure other things to a slab but don't provide any type of reinforcement. Choice *D* is wrong because pipe would not be a good reinforcement due to the hollow design.

21. C: Lathes are used to turn down metal or wood. This means that as the object is rotated, some material is removed to either reshape or smooth the object. This is not the same as grinding because material is actually cut away, so Choice *A* is wrong. Even though it rotates, it is not used to drive screws, so Choice *B* is wrong. It also can't be used to bend angles, so Choice *D* is wrong.

22. A: Proper air ventilation is always needed when welding. An exhaust extraction system should also be used to help remove the excess fumes. Choice *B* is wrong because there is no need for a hard hat when welding. However, a proper welding mask should be worn. Insulated clothing may be worn while welding but it is not necessary. Finally, choice *D* is wrong because a welding mask should be worn instead of safety glasses. The mask will protect the eyes as well as the entire face.

23. C: Shingles should be added to the roof from the bottom to the top. This ensures that there are no gaps for water to trickle through as runs down the roof. Adding them from the top to the bottom would create multiple gaps for water to run through, so Choice *D* is wrong. Trying to add them from either side would also make it difficult to get them to overlap correctly, so Choices *A* and *B* are also wrong.

24. D: A trowel is used to smooth and wipe away extra mortar. Mortar is used to hold the bricks together, not cement or plaster, so Choices *A* and *C* are wrong. A trommel is a piece of machinery that is used to separate materials, such as minerals and solid waste, so that means Choice *B* is also wrong.

25. B: MIG welders use a continuously feeding wire, while TIG welders use a welding rod. TIG welders also use a tungsten electrode to transfer the current to the welding arc, which means that Choice *A* is wrong. Choice *C* is wrong because TIG welders are the ones used for welding thinner gauged metal. Finally, Choice *D* is wrong because the room should always be well-ventilated, and have an exhaust fan, when welding.

Mechanical Comprehension

The **Mechanical Comprehension** (MC) section tests a candidate's knowledge of mechanics and physical principles. These include concepts of force, energy, and work, and how they're used to predict the functioning of tools and machines. This knowledge is important for a successful career in the military. A good score on the MC test shows that a candidate has a solid background for learning how to use tools and machines properly. This is extremely important for the efficient, safe completion of most tasks a future soldier, sailor, or airman must undertake during their service.

The test problems in the MC section of the exam focus on understanding physical principles, but they are *qualitative* in nature rather than *quantitative*. This means the problems involve predicting the *behavior* of a system (such as the direction it moves) rather than calculating a specific measurement (such as its velocity). The figure below shows a sample problem similar to those on the MC test:

Mechanical Comprehension Sample Test Problem

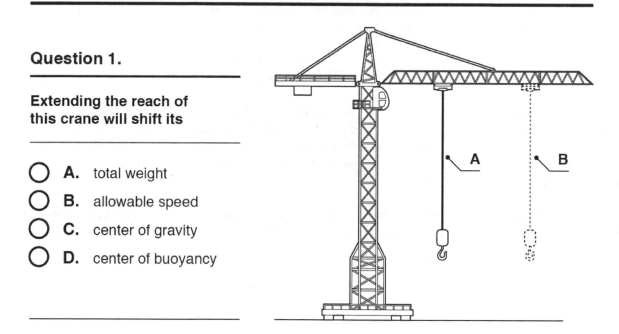

Question 1.

Extending the reach of this crane will shift its

- ○ **A.** total weight
- ○ **B.** allowable speed
- ○ **C.** center of gravity
- ○ **D.** center of buoyancy

The sample problem pictures a system of a crane lifting a weight, and below the picture is a question. On the exam, it's very important to read these questions carefully. This question involves completing the following sentence: *Extending the reach of this crane will shift its _____.* After the sentence, four possible answers are provided.

The correct answer is *C, center of gravity*. In this sample problem, it's easy to guess the correct answer simply by eliminating the rest. Choice A is incorrect because moving the load out along the crane's boom won't change its weight, just like moving a bodybuilder's arm that's holding a dumbbell won't change the combined weight of the bodybuilder and the dumbbell. Choice B is incorrect because the crane isn't moving. That leaves Choices *C* and *D*, but *D* is incorrect because buoyancy is only involved in systems with a liquid (the buoyancy of air is negligible). Therefore, through the process of elimination, *C* is the correct answer.

Review of Physics and Mechanical Principles

The proper use of tools and machinery depends on an understanding of basic physics, which includes the study of motion and the interactions of mass, force, and energy. These terms are used every day, but their exact meanings are difficult to define. In fact, they're usually defined in terms of each other.

The matter in the universe (atoms and molecules) is characterized in terms of its **mass**, which is measured in kilograms in the **International System of Units** (**SI**). The amount of mass that occupies a given volume of space is termed **density**.

Mass occupies space, but it's also a component that inversely relates to acceleration when a force is applied to it. This **force** is the application of energy to an object with the intent of changing its position (mainly its acceleration).

To understand **acceleration**, it's necessary to relate it to displacement and velocity. The **displacement** of an object is simply the distance it travels. The **velocity** of an object is the distance it travels in a unit of time, such as miles per hour or meters per second:

$$Velocity = \frac{Distance\ Traveled}{Time\ Required}$$

There's often confusion between the words "speed" and "velocity." Velocity includes speed *and* direction. For example, a car traveling east and another traveling west can have the same speed of 30 miles per hour (mph), but their velocities are different. If movement eastward is considered positive, then movement westward is negative. Thus, the eastbound car has a velocity of 30 mph while the westbound car has a velocity of -30 mph.

The fact that velocity has a **magnitude** (speed) and a direction makes it a vector quantity. A **vector** is an arrow pointing in the direction of motion, with its length proportional to its magnitude.

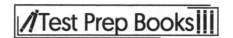

Vectors can be added geometrically as shown below. In this example, a boat is traveling east at 4 knots (nautical miles per hour) and there's a current of 3 knots (thus a slow boat and a very fast current). If the boat travels in the same direction as the current, it gets a "lift" from the current and its speed is 7 knots. If the boat heads *into* the current, it has a forward speed of only 1 knot (4 knots – 3 knots = 1 knot) and makes very little headway. As shown in the figure below, the current is flowing north across the boat's path. Thus, for every 4 miles of progress the boat makes eastward, it drifts 3 miles to the north.

Working with Velocity Vectors

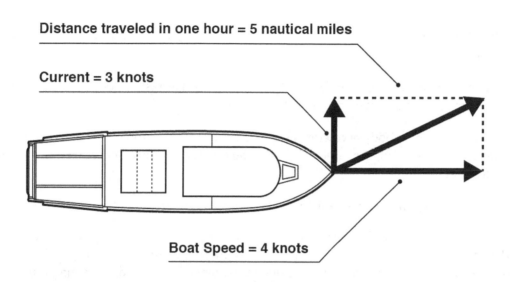

Distance traveled in one hour = 5 nautical miles

Current = 3 knots

Boat Speed = 4 knots

The total distance traveled is calculated using the *Pythagorean Theorem* for a right triangle, which should be memorized as follows:

$$a^2 + b^2 = c^2 \text{ or } c = \sqrt{a^2 + b^2}$$

Of course, the problem above was set up using a Pythagorean triple (3, 4, 5), which made the calculation easy.

Another example where velocity and speed are different is with a car traveling around a bend in the road. The speed is constant along the road, but the direction (and therefore the velocity) changes continuously.

The **acceleration** of an object is the change in its velocity in a given period of time:

$$Acceleration = \frac{Change\ in\ Velocity}{Time\ Required}$$

For example, a car starts at rest and then reaches a velocity of 70 mph in 8 seconds. What's the car's acceleration in feet per second squared? First, the velocity must be converted from miles per hour to feet per second:

$$70\frac{miles}{hour} \times \frac{5,280\ feet}{mile} \times \frac{hour}{3600\ seconds} = 102.67\ feet/second$$

Starting from rest, the acceleration is:

$$Acceleration = \frac{102.67\frac{feet}{second} - 0\frac{feet}{second}}{8\ seconds} = 12.8\ feet/second^2$$

Newton's Laws

Isaac Newton's three laws of motion describe how the acceleration of an object is related to its mass and the forces acting on it. The three laws are:

- Unless acted on by a force, a body at rest tends to remain at rest; a body in motion tends to remain in motion with a constant velocity and direction.

- A force that acts on a body accelerates it in the direction of the force. The larger the force, the greater the acceleration; the larger the mass, the greater its inertia (resistance to movement and acceleration).

- Every force acting on a body is resisted by an equal and opposite force.

To understand Newton's laws, it's necessary to understand forces. These forces can push or pull on a mass, and they have a magnitude and a direction. Forces are represented by a vector, which is the arrow lined up along the direction of the force with its tip at the point of application. The magnitude of the force is represented by the length of the vector.

The figure below shows a mass acted on or "pushed" by two equal forces (shown here by vectors of the same length). Both vectors "push" along the same line through the center of the mass, but in opposite directions. What happens?

A Mass Acted on by Equal and Opposite Forces

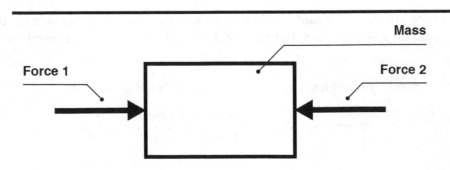

According to Newton's third law, every force on a body is resisted by an equal and opposite force. In the figure above, Force 1 acts on the left side of the mass. The mass pushes back. Force 2 acts on the right side, and the mass pushes back against this force too. The net force on the mass is zero, so according to Newton's first law, there's no change in the **momentum** (the mass times its velocity) of the mass. Therefore, if the mass is at rest before the forces are applied, it remains at rest. If the mass is in motion with a constant velocity, its momentum doesn't change. So, what happens when the net force on the mass isn't zero, as shown in the figure below?

A Mass Acted on by Unbalanced Forces

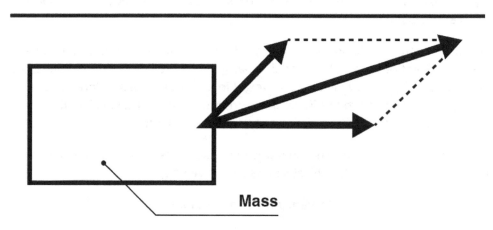

Mass

Notice that the forces are vector quantities and are added geometrically the same way that velocity vectors are manipulated.

Here in the figure above, the mass is pulled by two forces acting to the right, so the mass accelerates in the direction of the net force. This is described by Newton's second law:

Force = Mass x Acceleration

The force (measured in *newtons*) is equal to the product of the mass (measured in kilograms) and its acceleration (measured in meters per second squared or meters per second, per second). A better way to look at the equation is dividing through by the mass:

Acceleration = Force/Mass

This form of the equation makes it easier to see that the acceleration of an object varies directly with the net force applied and inversely with the mass. Thus, as the mass increases, the acceleration is reduced for a given force. To better understand, think of how a baseball accelerates when hit by a bat. Now imagine hitting a cannonball with the same bat and the same force. The cannonball is more massive than the baseball, so it won't accelerate very much when hit by the bat.

In addition to forces acting on a body by touching it, gravity acts as a force at a distance and causes all bodies in the universe to attract each other. The **force of gravity** (F_g) is proportional to the masses of the two objects (*m* and *M*) and inversely proportional to the square of the distance (r^2) between them (and *G* is the proportionality constant).

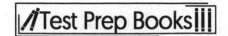

This is shown in the following equation:

$$F_g = G \frac{mM}{r^2}$$

The force of gravity is what causes an object to fall to Earth when dropped from an airplane. Understanding gravity helps explain the difference between mass and weight. Mass is a property of an object that remains the same while it's intact, no matter where it's located. A 10-kilogram cannonball has the same mass on Earth as it does on the moon. On Earth, it weighs 98.1 newtons because of the attractive force of gravity, so it accelerates at 9.81 m/s². However, on the moon, the same cannonball has a weight of only about 16 newtons. This is because the gravitational attraction on the moon is approximately one-sixth that on Earth. Although Earth still attracts the body on the moon, it's so far away that its force is negligible.

For Americans, there's often confusion when talking about mass because the United States still uses **pounds** as a measurement of weight. In the traditional system used in the United States, the unit of mass is called a **slug**. It's derived by dividing the weight in pounds by the acceleration of gravity (32 feet/s²); however, it's rarely used today. To avoid future problems, test takers should continue using SI units and remember to express mass in kilograms and weight in Newtons.

Another way to understand Newton's second law is to think of it as an object's change in momentum, which is defined as the product of the object's mass and its velocity:

Momentum = Mass x Velocity

Which of the following has the greater momentum: a pitched baseball, a softball, or a bullet fired from a rifle?

A bullet with a mass of 5 grams (0.005 kilograms) is fired from a rifle with a muzzle velocity of 2200 mph. Its momentum is calculated as:

$$2200 \frac{miles}{hour} \times \frac{5{,}280\ feet}{mile} \times \frac{m}{3.28\ feet} \times \frac{hour}{3600\ seconds} \times 0.005kg = 4.92 \frac{kg.m}{seconds}$$

A softball has a mass between 177 grams and 198 grams and is thrown by a college pitcher at 50 miles per hour. Taking an average mass of 188 grams (0.188 kilograms), a softball's momentum is calculated as:

$$50 \frac{miles}{hour} \times \frac{5280\ feet}{mile} \times \frac{m}{3.28\ ft} \times \frac{hour}{3600\ seconds} \times 0.188kg = 4.19 \frac{kg.m}{seconds}$$

That's only slightly less than the momentum of the bullet. Although the speed of the softball is considerably less, its mass is much greater than the bullet's.

A professional baseball pitcher can throw a 145-gram baseball at 100 miles per hour. A similar calculation (try doing it!) shows that the pitched hardball has a momentum of about 6.48 kg.m/seconds. That's more momentum than a speeding bullet!

So why is the bullet more harmful than the hard ball? It's because the force that it applies acts on a much smaller area.

Instead of using acceleration, Newton's second law is expressed here as the change in momentum (with the delta symbol "Δ" meaning "change"):

$$Force = \frac{\Delta\,Momentum}{\Delta\,Time} = \frac{\Delta\,(Mass\,\times\,Velocity)}{\Delta\,Time} = Mass\,\times\,\frac{\Delta\,Velocity}{\Delta\,Time}$$

The rapid application of force is called **impulse.** Another way of stating Newton's second law is in terms of the impulse, which is the force multiplied by its time of application:

$$Impluse = Force \times \Delta\,Time = Mass \times \Delta\,Velocity$$

In the case of the rifle, the force created by the pressure of the charge's explosion in its shell pushes the bullet, accelerating it until it leaves the barrel of the gun with its **muzzle velocity** (the speed the bullet has when it leaves the muzzle). After leaving the gun, the bullet doesn't accelerate because the gas pressure is exhausted. The bullet travels with a constant velocity in the direction it's fired (ignoring the force exerted against the bullet by friction and drag).

Similarly, the pitcher applies a force to the ball by using their muscles when throwing. Once the ball leaves the pitcher's fingers, it doesn't accelerate and the ball travels toward the batter at a constant speed (again ignoring friction and drag). The speed is constant, but the velocity can change if the ball travels along a curve.

Projectile Motion

According to Newton's first law, if no additional forces act on the bullet or ball, it travels in a straight line. This is also true if the bullet is fired in outer space. However, here on Earth, the force of gravity continues to act so the motion of the bullet or ball is affected.

What happens when a bullet is fired from the top of a hill using a rifle held perfectly horizontal? Ignoring air resistance, its horizontal velocity remains constant at its muzzle velocity. Its vertical velocity (which is zero when it leaves the gun barrel) increases because of gravity's acceleration. Each passing second, the bullet traces out the same distance horizontally while increasing distance vertically (shown in the figure below). In the end, the projectile traces out a **parabolic curve**.

Projectile Path for a Bullet Fired Horizontally from a Hill (Ignoring Air Resistance)

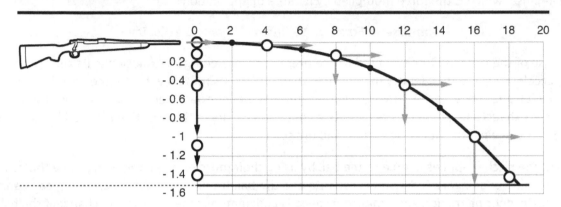

This vertical, downward acceleration is why a pitcher must put an arc on the ball when throwing across home plate. Otherwise the ball will fall at the batter's feet.

It's also interesting to note that if an artillery crew simultaneously drops one cannonball and fires another one horizontally, the two cannonballs will hit the ground at the same time since both balls are accelerating at the same rate and experience the same changes in vertical velocity.

What if air resistance is taken into account? This is best answered by looking at the horizontal and vertical motions separately.

The horizontal velocity is no longer constant because the initial velocity of the projectile is continually reduced by the resistance of the air. This is a complex problem in fluid mechanics, but it's sufficient to note that that the projectile doesn't fly as far before landing as predicted from the simple theory.

The vertical velocity is also reduced by air resistance. However, unlike the horizontal motion where the propelling force is zero after the cannonball is fired, the downward force of gravity acts continuously. The downward velocity increases every second due to the acceleration of gravity. As the velocity increases, the resisting force (called **drag**) increases with the square of the velocity. If the projectile is fired or dropped from a sufficient height, it reaches a terminal velocity such that the upward drag force equals the downward force of gravity. When that occurs, the projectile falls at a constant rate.

This is the same principle that's used for a parachute. Its drag (caused by its shape that scoops up air) is sufficient enough to slow down the fall of the parachutist to a safe velocity, thus avoiding a fatal crash on the ground.

So, what's the bottom line? If the vertical height isn't too great, a real projectile will fall short of the theoretical point of impact. However, if the height of the fall is significant and the drag of the object results in a small terminal fall velocity, then the projectile can go further than the theoretical point of impact.

What if the projectile is launched from a moving platform? In this case, the platform's velocity is added to the projectile's velocity. That's why an object dropped from the mast of a moving ship lands at the base of the mast rather than behind it. However, to an observer on the shore, the object traces out a parabolic arc.

Angular Momentum

In the previous examples, all forces acted through the center of the mass, but what happens if the forces aren't applied through the same line of action, like in the figure below?

A Mass Acted on by Forces Out of Line with Each Other

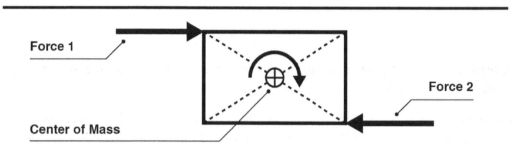

When this happens, the two forces create **torque** and the mass rotates around its center of gravity. In the figure above, the center of gravity is the center of the rectangle (**Center of Mass**), which is determined by the two, intersecting main diagonals. The center of an irregularly shaped object is found by hanging it from two different edges, and the center of gravity is at the intersection of the two "plumb lines."

Newton's second law still applies when the forces form a moment pair, but it must be expressed in terms of angular acceleration and the moment of inertia. The **moment of inertia** is a measure of the body's resistance to rotation, similar to the mass's resistance to linear acceleration. The more compact the body, the less the moment of inertia and the faster it rotates, much like how an ice skater spinning with outstretched arms will speed up as the arms are brought in close to the body.

The concept of torque is important in understanding the use of wrenches and is likely to be on the test. The concept of torque and moment/lever arm will be taken up again below when the physics of simple machines is presented.

Energy and Work

The previous examples of moving boats, cars, bullets, and baseballs are examples of simple systems that are thought of as particles with forces acting through their center of gravity. They all have one property in common: **energy**. The energy of the system results from the forces acting on it and is considered its ability to do work.

Work or the energy required to do work (which are the same) is calculated as the product of force and distance traveled along the line of action of the force. It's measured in **foot-pounds** in the traditional system (which is still used in workshops and factories) and in **newton meters** ($N \cdot m$) in the International System of Units (SI), which is the preferred system of measurement today.

Potential and Kinetic Energy

Energy can neither be created nor destroyed, but it can be converted from one form to another. There are many forms of energy, but it's useful to start with mechanical energy and potential energy.

The **potential energy** of an object is equal to the work that's required to lift it from its original elevation to its current elevation. This is calculated as the weight of the object or its downward force (mass times the acceleration of gravity) multiplied by the distance (y) it is lifted above the reference elevation or "datum." This is written:

$$PE = mgy$$

The mechanical or **kinetic energy** of a system is related to its mass and velocity:

$$KE = \frac{1}{2}mv^2$$

The **total energy** is the sum of the kinetic energy and the potential energy, both of which are measured in foot-pounds or newton meters.

If a weight with a mass of 10 kilograms is raised up a ladder to a height of 10 meters, it has a potential energy of 10m x 10kg x 9.81m/s^2 = 981N·m. This is approximately 1000 newton meters if the acceleration of gravity (9.81 m/s^2) is rounded to 10 m/s^2, which is accurate enough for most earth-bound calculations. It has zero kinetic energy because it's at rest, with zero velocity.

If the weight is dropped from its perch, it accelerates downward so that its velocity and kinetic energy increase as its potential energy is "used up" or, more precisely, converted to kinetic energy.

When the weight reaches the bottom of the ladder, just before it hits the ground, it has a kinetic energy of 981 N·m (ignoring small losses due to air resistance). The velocity can be solved by using the following:

$$981\ N \cdot m = \frac{1}{2} 10\ kg \times v^2 \quad or \quad v = 14.01\ m/s$$

When the 10-kilogram weight hits the ground, its potential energy (which was measured from the ground) and its velocity are both zero, so its kinetic energy is also zero. What's happened to the energy? It's dissipated into heat, noise, and kicking up some dust. It's important to remember that energy can neither be created nor destroyed, so it can only change from one form to another.

The conversion between potential and kinetic energy works the same way for a pendulum. If it's raised and held at its highest position, it has maximum potential energy but zero kinetic energy.

Potential and Kinetic Energy for a Swinging Pendulum

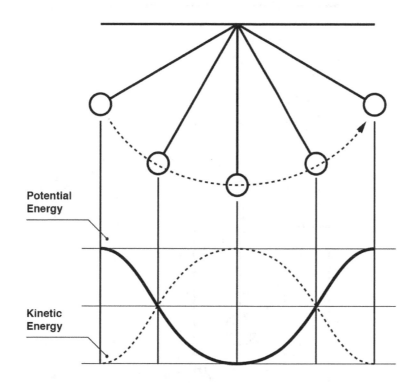

When the pendulum is released from its highest position (see left side of the figure above), it swings down so that its kinetic energy increases as its potential energy decreases. At the bottom of its swing, the pendulum is moving at its maximum velocity with its maximum kinetic energy. As the pendulum swings past the bottom of its path, its velocity slows down as its potential energy increases.

Work
The released potential energy of a system can be used to do **work**.

For instance, most of the energy lost by letting a weight fall freely can be recovered by hooking it up to a pulley to do work by pulling another weight back up (as shown in the figure below).

Using the Energy of a Falling Weight to Raise Another Weight

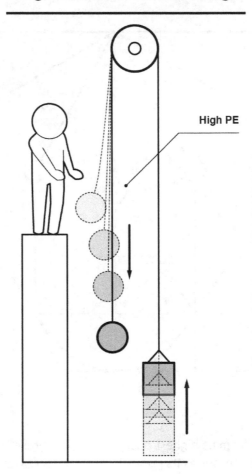

High PE

In other words, the potential energy expended to lower the weight is used to do the work of lifting another object. Of course, in a real system, there are losses due to friction. The action of pulleys will be discussed later in this study guide.

Since **energy** is defined as *the capacity to do work*, energy and work are measured in the same units:

$$Energy = Work = Force \times Distance$$

Force is measured in **newtons** (*N*)**.** Distance is measured in meters. The units of work are **newton meters** *(N·m)*. The same is true for kinetic energy and potential energy.

Another way to store energy is to compress a spring. Energy is stored in the spring by stretching or compressing it. The work required to shorten or lengthen the spring is given by the equation:

$$F = k \times d$$

Here, "d" is the length in meters and "k" is the resistance of the spring constant (measured in N·m), which is a constant as long as the spring isn't stretched past its elastic limit. The resistance of the spring is constant, but the force needed to compress the spring increases with each millimeter it's pushed.

The potential energy stored in the spring is equal to the work done to compress it, which is the total force times the change in length. Since the resisting force of the spring increases as its displacement increases, the average force must be used in the calculation:

$$W = PE = F \times d = \frac{1}{2}(F_i + F_f)d \times d$$

$$\frac{1}{2}(0 + F_f)d \times d = \frac{1}{2}Fd^2$$

The potential energy in the spring is stored by locking it into place, and the work energy used to compress it is recovered when the spring is unlocked. It's the same when dropping a weight from a height—the energy doesn't have to be wasted. In the case of the spring, the energy is used to propel an object.

Potential and Kinetic Energy of a Spring

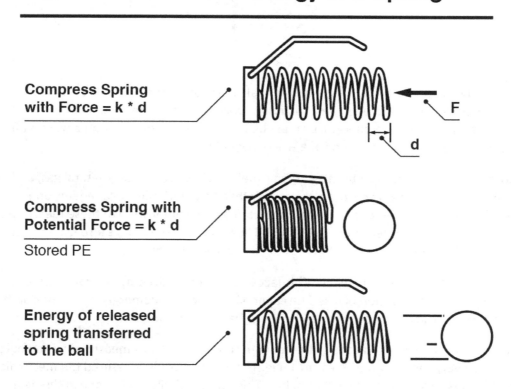

Pushing a block horizontally along a rough surface requires work. In this example, the work needs to overcome the force of friction, which opposes the direction of the motion and equals the weight of the block times a **friction factor** (*f*). The friction factor is greater for rough surfaces than smooth surfaces,

and it's usually greater *before* the motion starts than after it has begun to slide. These terms are illustrated in the figure below.

Pushing a Block Horizontally
Against the Force of Friction

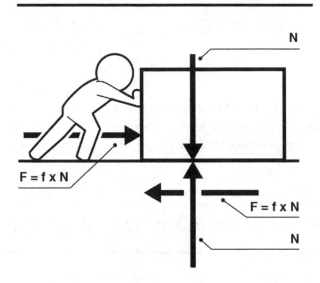

When pushing a block, there's no increase in potential energy since the block's elevation doesn't change. Expending the energy to overcome friction is "wasted" in the generation of heat. Yet, to move a block from point A to point B, an energy cost must be paid. However, friction isn't always a hindrance. In fact, it's the force that makes the motion of a wheel possible.

Heat energy can also be created by burning organic fuels, such as wood, coal, natural gas, and petroleum. All of these are derived from plant matter that's created using solar energy and photosynthesis. The chemical energy or *"heat"* liberated by the combustion of these fuels is used to warm buildings during the winter or even melt metal in a foundry. The heat is also used to generate steam, which can drive engines or turn turbines to generate electric energy.

In fact, work and heat are interchangeable. This fact was first recognized by gun founders when they were boring out cast, brass cannon blanks. The cannon blanks were submerged in a water bath to reduce friction, yet as the boring continued, the water bath boiled away!

Later, the amount of work needed to raise the temperature of water was measured by an English physicist (and brewer) named James Prescott Joule. The way that Joule measured the mechanical equivalent of heat is illustrated in the figure below. This setup is similar to the one in the figure above with the pulley, except instead of lifting another weight, the falling weight's potential energy is converted to the mechanical energy of the rotating vertical shaft. This turns the paddles, which churns the water to increase its temperature. Through a long series of repeated measurements, Joule showed

that 4186 N·m of work was necessary to raise the temperature of one kilogram of water by one degree Celsius, no matter how the work was delivered.

Device Measuring the Mechanical Energy Needed to Increase the Temperature of Water

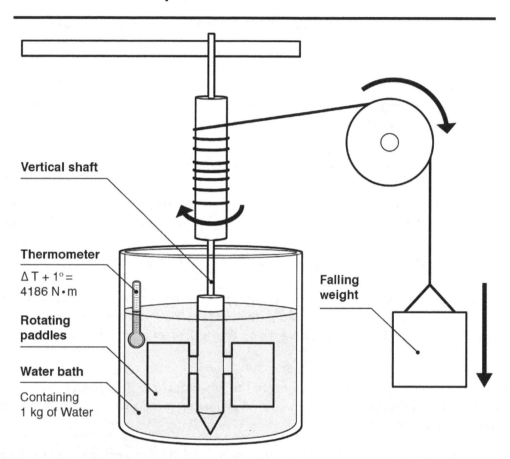

Vertical shaft

Thermometer

$\Delta T + 1° = 4186\ N \cdot m$

Rotating paddles

Water bath

Containing 1 kg of Water

Falling weight

In recognition of this experiment, the newton meter is also called a **joule**. Linking the names for work and heat to the names of two great physicists is truly appropriate because heat and work being interchangeable is of the greatest practical importance. These two men were part of a very small, select group of scientists for whom units of measurement have been named: Marie Curie for radioactivity, Blaise Pascal for pressure, James Watt for power, Andre Ampere for electric current, and only a few others.

Just as mechanical work is converted into heat energy, heat energy is converted into mechanical energy in the reverse process. An example of this is a closely fitting piston supporting a weight and mounted in a cylinder where steam enters from the bottom.

In this example, water is heated into steam in a boiler, and then the steam is drawn off and piped into a cylinder. Steam pressure builds up in the piston, exerting a force in all directions. This is counteracted by the tensile strength of the cylinder; otherwise, it would burst. The pressure also acts on the exposed face of the piston, pushing it upwards against the load (displacing it) and thus doing work.

Work developed from the pressure acting over the area exerts a force on the piston as described in the following equation:

$$Work = Pressure \times Piston\ Area \times Displacement$$

Here, the work is measured in newton meters, the pressure in newtons per square meter or **pascals** (*Pa*), and the piston displacement is measured in meters.

Since the volume enclosed between the cylinder and piston increases with the displacement, the work can also be expressed as:

$$Work = Pressure \times \Delta\ Volume$$

For example, a 10-kilogram weight is set on top of a piston-cylinder assembly with a diameter of 25 centimeters. The area of the cylinder is:

$$Area = \frac{\pi \times d^2}{4} = 0.785 \times 0.25^2 = .049\ m^2$$

If the acceleration due to gravity is approximately 10 m/s², and the area is rounded to .05 meters squared, then the pressure needed to counteract the weight of the 10-kilogram weight is estimated as:

$$P = \frac{F}{A} \approx 10\ \times \frac{10}{0.05} = 2000\ \frac{N}{m^2} = 2000\ Pa = 2\ KPa$$

If steam with a pressure slightly greater than this value is piped into the cylinder, it slowly lifts the load. If steam at a much higher pressure is suddenly admitted to the cylinder, it throws the load into the air. This is the principle used to steam-catapult airplanes off the deck of an aircraft carrier.

Power

Power is defined as the rate at which work is done, or the time it takes to do a given amount of work. In the International System of Units (SI), work is measured in **newton meters** (*N·m*) or **joules** (*J*). Power is measured in joules/second or *watts (W)*.

For example, to raise a 1-kilogram mass one meter off the ground, it takes approximately 10 newton meters of work (approximating the gravitational acceleration of 9.81 m/s² as 10 m/s²). To do the work in 10 seconds, it requires 1 watt of power. Doing it in 1 second requires 10 watts of power. Essentially, *doing it faster means dividing by a smaller number*, and that means greater power.

Although SI units are preferred for technical work throughout the world, the old traditional (or English) unit of measuring power is still used. Introduced by James Watt (the same man for whom the SI unit of power "watt" is named), the unit of **horsepower** (*HP*) rated the power of the steam engines that he and his partner (Matthew Boulton) manufactured and sold to mine operators in 18th century England. The mine operators used these engines to pump water out of flooded facilities in the beginning of the Industrial Revolution.

To provide a measurement that the miners would be familiar with, Watt and Boulton referenced the power of their engines with the "power of a horse."

Watt's measurements showed that, on average, a well-harnessed horse could lift a 330-pound weight 100 feet up a well in one minute (330 pounds is the weight of a 40-gallon barrel filled to the brim).

Remembering that power is expressed in terms of energy or work per unit time, horsepower came to be measured as:

$$1\ HP = \frac{100\ feet\ \times\ 330\ pounds}{1\ minute} \times \frac{1\ minute}{60\ seconds} = 550\ foot\ pounds/second$$

A horse that pulled the weight up faster, or pulled up more weight in the same time, was a more *powerful* horse than Watt's "average horse."

Hundreds of millions of engines of all types have been built since Watt and Boulton started manufacturing their products, and the unit of horsepower has been used throughout the world to this day. Of course, modern technicians and engineers still need to convert horsepower to watts to work with SI units. An approximate conversion is *1 HP = 746 W*.

Take for example a 2016 CTS-V Cadillac rated at 640 HP. If a **megawatt** (*MW*) is one million watts, that means the Cadillac has almost half a megawatt of power as shown by this conversion:

$$640\ HP \times \frac{746\ W}{1\ HP} = 477,440\ W = 477.4\ kW$$

The power of the Cadillac is comparable to that of the new Westinghouse AP-1000 Nuclear Power Plant, which is rated at 1100 MW or the equivalent of 2304 Cadillacs (assuming no loss in power). That would need a very big parking lot and a tremendous amount of gasoline!

A question that's often asked is, "How much energy is expended by running an engine for a fixed amount of time?" This is important to know when planning how much fuel is needed to run an engine. For example, how much energy is expended in running the new Cadillac at maximum power for 30 minutes?

In this case, the energy expenditure is approximately 240 kilowatt hours. This must be converted to joules, using the conversion factor that one watt equals one joule per second:

$$240,000\ W\ hours\ \times \frac{3600\ seconds}{1\ hour} = 8.64(10)^8\ joules$$

So how much gasoline is burned? Industrial tests show that a gallon of gasoline is rated to contain about 1.3×10^8 joules of energy. That's 130 million joules per gallon. The gallons of gasoline are obtained by dividing:

$$\frac{8.64(10)^8 J}{1.3(10)^8 J/gallon} = 6.65\ gallons\ \times \frac{3.8\ liters}{gallon} = 25.3\ liters$$

The calculation has now come full circle. It began with power. Power equals energy divided by time. Power multiplied by time equals the energy needed to run the machine, which came from burning fuel.

Fluids

In addition to the behavior of solid particles acted on by forces, it is important to understand the behavior of fluids. **Fluids** include both liquids and gasses. The best way to understand fluid behavior is to contrast it with the behavior of solids, as shown in the figure below.

First, consider a block of ice, which is solid water. If it is set down inside a large box it will exert a force on the bottom of the box due to its weight as shown on the left, in Part A of the figure. The solid block exerts a pressure on the bottom of the box equal to its total weight divided by the area of its base:

$$Pressure = Weight\ of\ block/Area\ of\ base$$

That pressure acts only in the area directly under the block of ice.

If the same mass of ice is melted, it behaves much differently. It still has the same weight as before because its mass hasn't changed. However, the volume has decreased because liquid water molecules are more tightly packed together than ice molecules, which is why ice floats (it is less dense).

The Behavior of Solids and Liquids Compared

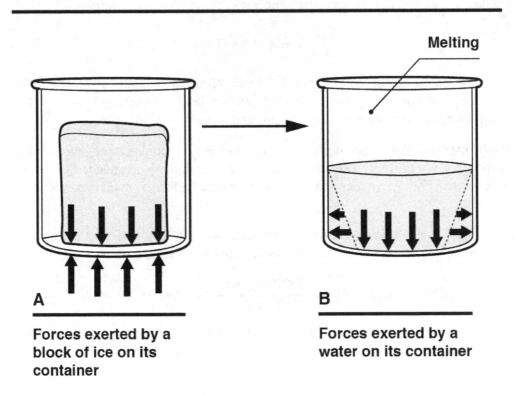

A

Forces exerted by a block of ice on its container

B

Forces exerted by a water on its container

The melted ice (now water) conforms to the shape of the container. This means that the fluid exerts pressure not only on the base, but on the sides of the box at the water line and below. Actually, pressure in a liquid is exerted in all directions, but all the forces in the interior of the fluid cancel each other out, so that a net force is only exerted on the walls. Note also that the pressure on the walls increases with the depth of the water.

The fact that the liquid exerts pressure in all directions is part of the reason some solids float in liquids. Consider the forces acting on a block of wood floating in water, as shown in the figure below.

Floatation of a Block of Wood

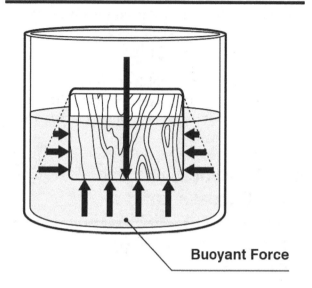

Buoyant Force

The block of wood is submerged in the water and pressure acts on its bottom and sides as shown. The weight of the block tends to force it down into the water. The force of the pressure on the left side of the block just cancels the force of the pressure on the right side.

There is a net upward force on the bottom of the block due to the pressure of the water acting on that surface. This force, which counteracts the weight of the block, is known as the **buoyant force**.

The block will sink to a depth such that the buoyant force of the water (equal to the weight of the volume displaced) just matches the total weight of the block. This will happen if two conditions are met:

- The body of water is deep enough to float the block
- The density of the block is less than the density of the water

If the body of water is not deep enough, the water pressure on the bottom side of the block won't be enough to develop a buoyant force equal to the block's weight. The block will be "beached" just like a boat caught at low tide.

If the density of the block is greater than the density of the fluid, the buoyant force acting on the bottom of the boat will not be sufficient to counteract the total weight of the block. That's why a solid steel block will sink in water.

If steel is denser than water, how can a steel ship float? The steel ship floats because it's hollow. The volume of water displaced by its steel shell (hull) is heavier than the entire weight of the ship and its contents (which includes a lot of empty space). In fact, there's so much empty space within a steel ship's hull that it can bob out of the water and be unstable at sea if some of the void spaces (called ballast tanks) aren't filled with water. This provides more weight and balance (or "trim") to the vessel.

The discussion of buoyant forces on solids holds for liquids as well. A less dense liquid can float on a denser liquid if they're **immiscible** (do not mix). For instance, oil can float on water because oil isn't as dense as the water. Fresh water can float on salt water for the same reason.

Pascal's law states that a change in pressure, applied to an enclosed fluid, is transmitted undiminished to every portion of the fluid and to the walls of its containing vessel. This principle is used in the design of hydraulic jacks, as shown in the figure below.

A force (F_1) is exerted on a small "driving" piston, which creates pressure on the hydraulic fluid. This pressure is transmitted through the fluid to a large cylinder. While the pressure is the same everywhere in the oil, the pressure action on the area of the larger cylinder creates a much higher upward force (F_2).

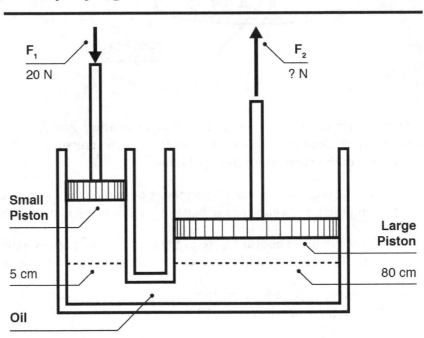

Illustration of a Hydraulic Jack Exemplifying Pascal's Law

Looking again at the figure above, suppose the diameter of the small cylinder is 5 centimeters and the diameter of the large cylinder is 80 centimeters. If a force of 20 newtons (N) is exerted on the small driving piston, what's the value of the upward force F_2? In other words, what weight can the large piston support?

The pressure within the system is created from the force F_1 acting over the area of the piston:

$$P = \frac{F_1}{A} = \frac{20\ N}{\pi\ (0.05\ m)^2/4} = 10,185\ Pa$$

The same pressure acts on the larger piston, creating the upward force, F_2:

$$F_2 = P \times A = 10{,}185\ Pa \times \pi \times (0.8\ m)^2/4 = 5120\ N$$

Because a liquid has no internal shear strength, it can be transported in a pipe or channel between two locations. A fluid's **rate of flow** is the volume of fluid that passes a given location in a given amount of time and is expressed in $m^3/second$. The **flow rate** (Q) is determined by measuring the **area of flow** (A) in m^2, and the **flow velocity** (v) in m/s:

$$Q = v \times A$$

This equation is called the **Continuity Equation**. It's one of the most important equations in engineering and should be memorized. For example, what is the flow rate for a pipe with an inside diameter of 1200 millimeters running full with a velocity of 1.6 m/s (measured by a **sonic velocity meter**)?

Using the Continuity Equation, the flow is obtained by keeping careful track of units:

$$Q = v \times A = 1.6\frac{m}{s} \times \frac{\pi}{4} \times \left(\frac{1200\ mm}{1000\ mm/m}\right)^2 = 1.81\ m^3/second$$

For more practice, imagine that a pipe is filling a storage tank with a diameter of 100 meters. How long does it take for the water level to rise by 2 meters?

Since the flow rate (Q) is expressed in m³/second, and volume is measured in m³, then the time in seconds to supply a volume (V) is V/Q. Here, the volume required is:

$$Volume\ Required = Base\ Area \times Depth = \frac{\pi}{4}100^2 \times 2\ m = 15{,}700\ m^3$$

Thus, the time to fill the tank another 2 meters is 15,700 m^3 divided by 1.81 m^3/s = 8674 seconds or 2.4 hours.

It's important to understand that, for a given flow rate, a smaller pipe requires a higher velocity.

The energy of a flow system is evaluated in terms of potential and kinetic energy, the same way the energy of a falling weight is evaluated. The total energy of a fluid flow system is divided into potential energy of elevation, and pressure and the kinetic energy of velocity. **Bernoulli's Equation** states that, for a constant flow rate, the total energy of the system (divided into components of elevation, pressure, and velocity) remains constant. This is written as:

$$Z + \frac{P}{\rho g} + \frac{v^2}{2g} = Constant$$

Each of the terms in this equation has dimensions of meters. The first term is the **elevation energy**, where Z is the elevation in meters. The second term is the **pressure energy**, where P is the pressure, ρ is the density, and g is the acceleration of gravity. The dimensions of the second term are also in meters. The third term is the **velocity energy**, also expressed in meters.

For a fixed elevation, the equation shows that, as the pressure increases, the velocity decreases. In the other case, as the velocity increases, the pressure decreases.

The use of the Bernoulli Equation is illustrated in the figure below. The total energy is the same at Sections 1 and 2. The area of flow at Section 1 is greater than the area at Section 2. Since the flow rate is the same at each section, the velocity at Point 2 is higher than at Point 1:

$$Q = V_1 \times A_1 = V_2 \times A_2, \qquad V_2 = V_1 \times \frac{A_1}{A_2}$$

Finally, since the total energy is the same at the two sections, the pressure at Point 2 is less than at Point 1. The tubes drawn at Points 1 and 2 would actually have the water levels shown in the figure; the pressure at each point would support a column of water of a height equal to the pressure divided by the unit weight of the water ($h = P/\rho g$).

An Example of Using the Bernoulli Equation

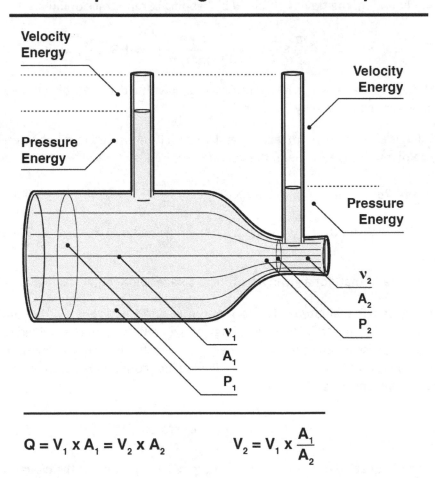

$$Q = V_1 \times A_1 = V_2 \times A_2 \qquad V_2 = V_1 \times \frac{A_1}{A_2}$$

Machines

Now that the basic physics of work and energy have been discussed, the common machines used to do the work can be discussed in more detail.

A **machine** is a device that: transforms energy from one form to another, multiplies the force applied to do work, changes the direction of the resultant force, or increases the speed at which the work is done.

The details of how energy is converted into work by a system are extremely complicated but, no matter how complicated the "linkage" between the components, every system is composed of certain elemental or simple machines. These are discussed briefly in the following sections.

Levers

The simplest machine is a **lever**, which consists of two pieces or components: a **bar** (or beam) and a **fulcrum** (the pivot-point around which motion takes place). As shown below, the **effort** acts at a distance (L_1) from the fulcrum and the **load** acts at a distance (L_2) from the fulcrum.

Components of a Lever

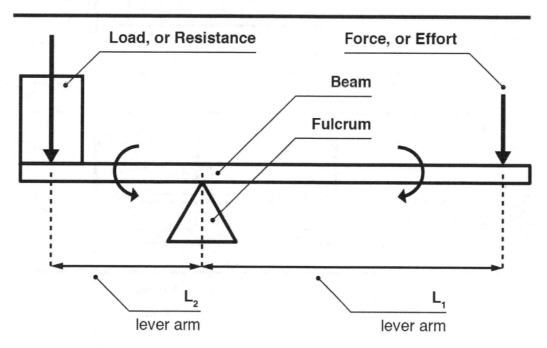

These lengths L_1 and L_2 are called **lever arms**. When the lever is balanced, the load (R) times its lever arm (L_2) equals the effort (F) times its lever arm (L_1). The force needed to lift the load is:

$$F = R \times \frac{L_2}{L_1}$$

This equation shows that as the lever arm L_1 is increased, the force required to lift the resisting load (R) is reduced. This is why Archimedes, one of the leading ancient Greek scientists, said, "Give me a lever long enough, and a place to stand, and I can move the Earth."

The ratio of the moment arms is the so-called "mechanical advantage" of the simple lever; the effort is multiplied by the mechanical advantage. For example, a 100-kilogram mass (a weight of approximately

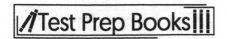

1000 N) is lifted with a lever like the one in the figure below, with a total length of 3 meters, and the fulcrum situated 50 centimeters from the left end. What's the force needed to balance the load?

$$F = 1000\ N\ \times \frac{0.5\ meters}{2.5\ meters} = 200\ N$$

Depending on the location of the load and effort with respect to the fulcrum, three "classes" of lever are recognized. In each case, the forces can be analyzed as described above.

The Three Classes of Levers

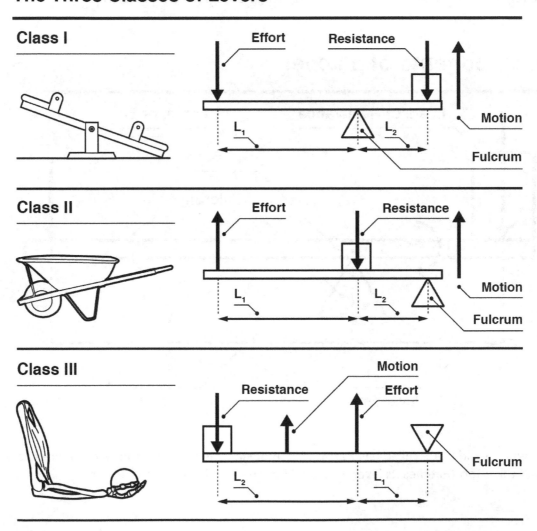

As seen in the figure, a **Class I** lever has the fulcrum positioned between the effort and the load. Examples of Class I levers include see-saws, balance scales, crow bars, and scissors. As explained above, the force needed to balance the load is $F = R \times (L_2/L_1)$, which means that the mechanical advantage is L_2/L_1. The crane boom shown back in the first figure in this section was a Class I lever, where the tower acted as the fulcrum and the counterweight on the left end of the boom provided the effort.

For a **Class II** lever, the load is placed between the fulcrum and the effort. A wheel barrow is a good example of a Class II lever. The mechanical advantage of a Class II lever is $(L_1 + L_2)/L_2$.

For a **Class III** lever, the effort is applied at a point between the fulcrum and the load, which increases the speed at which the load is moved. A human arm is a Class III lever, with the elbow acting as the fulcrum. The mechanical advantage of a Class III lever is $(L_1 + L_2)/L_1$.

Wheels and Axles

The wheel and axle is a special kind of lever. The **axle**, to which the load is applied, is set perpendicular to the **wheel** through its center. Effort is then applied along the rim of the wheel, either with a cable running around the perimeter or with a **crank** set parallel to the axle.

The mechanical advantage of the wheel and axle is provided by the moment arm of the perimeter cable or crank. Using the center of the axle (with a radius of r) as the fulcrum, the resistance of the load (L) is just balanced by the effort (F) times the wheel radius:

$$F \times R = L \times r \quad \text{or} \quad F = L \times \frac{r}{R}$$

This equation shows that increasing the wheel's radius for a given shaft reduces the required effort to carry the load. Of course, the axle must be made of a strong material or it'll be twisted apart by the applied torque. This is why steel axles are used.

Gears, Belts, and Cams

The functioning of a wheel and axle can be modified with the use of gears and belts. **Gears** are used to change the direction or speed of a wheel's motion.

The direction of a wheel's motion can be changed by using **beveled gears**, with the shafts set at right angles to each other, as shown in part A in the next figure.

The speed of a wheel can be changed by meshing together **spur gears** with different diameters. A small gear (A) is shown driving a larger gear (B) in the middle section (B) in the figure below. The gears rotate in opposite directions; if the driver, Gear A, moves clockwise, then Gear B is driven counter-clockwise. Gear B rotates at half the speed of the driver, Gear A. In general, the change in speed is given by the ratio of the number of teeth in each gear:

$$\frac{Rev_{Gear\ B}}{Rev_{Gear\ A}} = \frac{Number\ of\ Teeth\ in\ A}{Number\ of\ Teeth\ in\ B}$$

Rather than meshing the gears, **belts** are used to connect them as shown in part *(C)*.

Gear and Belt Arrangements

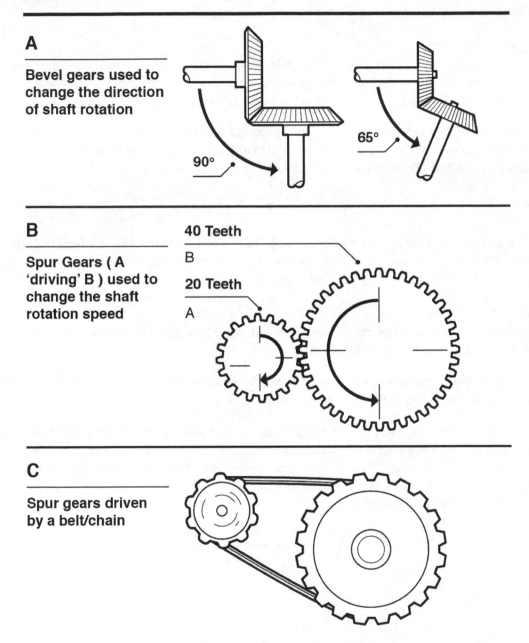

A

Bevel gears used to change the direction of shaft rotation

90°

65°

B

Spur Gears (A 'driving' B) used to change the shaft rotation speed

40 Teeth

B

20 Teeth

A

C

Spur gears driven by a belt/chain

Gears can change the speed and direction of the axle rotation, but the rotary motion is maintained. To convert the rotary motion of a gear train into linear motion, it's necessary to use a **cam** (a type of off-

centered wheel shown in the figure below, where rotary shaft motion lifts the valve in a vertical direction.

Conversion of Rotary to Vertical Linear Motion with a Cam

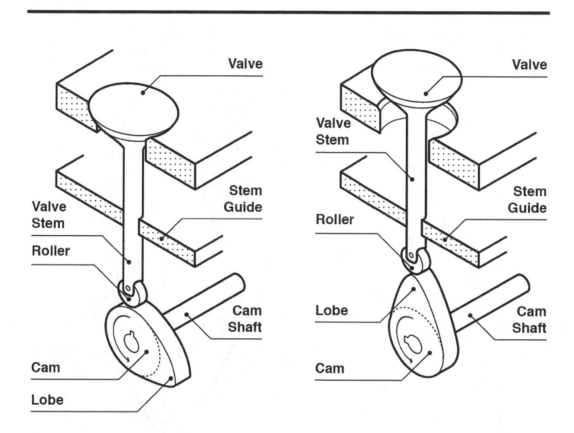

Pulleys

A **pulley** looks like a wheel and axle but provides a mechanical advantage in a different way. A **fixed pulley** was shown previously as a way to capture the potential energy of a falling weight to do useful work by lifting another weight. As shown in part A in the figure below, the fixed pulley is used to change the direction of the downward force exerted by the falling weight, but it doesn't provide any mechanical advantage.

The lever arm of the falling weight (A) is the distance between the rim of the fixed pulley and the center of the axle. This is also the length of the lever arm acting on the rising weight (B), so the ratio of the two arms is 1:0, meaning there's no mechanical advantage. In the case of a wheel and axle, the mechanical advantage is the ratio of the wheel radius to the axle radius.

A **moving pulley**, which is really a Class II lever, provides a mechanical advantage of 2:1 as shown below on the right side of the figure *(B)*.

Fixed-Block Versus Moving-Block Pulleys

A

Single Fixed Block with No Mechanical Advantage

B

Single Moving Block with 2:1 Mechanical Advantage

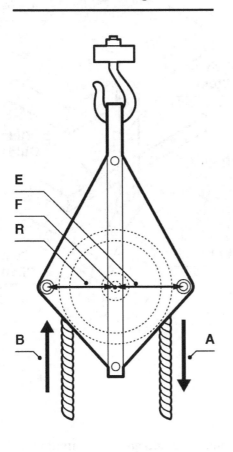

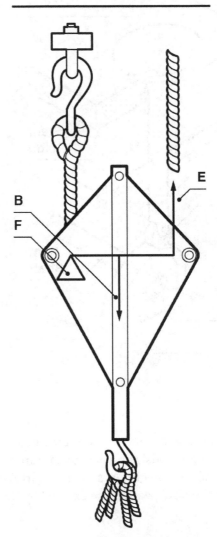

As demonstrated by the rigs in the figure below, using a wider moving block with multiple sheaves can achieve a greater mechanical advantage.

Single-Acting and Double-Acting Block and Tackles

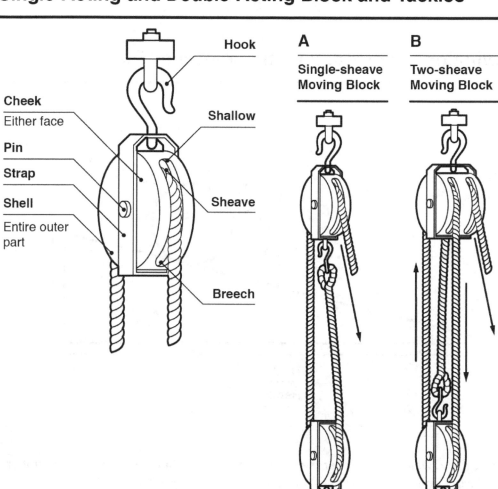

The mechanical advantage of the multiple-sheave block and tackle is approximated by counting the number of ropes going to and from the moving block. For example, there are two ropes connecting the moving block to the fixed block in part *A* of the figure above, so the mechanical advantage is 2:1. There are three ropes connecting the moving and fixed blocks in part *B*, so the mechanical advantage is 3:1. The advantage of using a multiple-sheave block is the increased hauling power obtained, but there's a cost; the weight of the moving block must be overcome, and a multiple-sheave block is significantly heavier than a single-sheave block.

Ramps

The **ramp** (or inclined plane) has been used since ancient times to move massive, extremely heavy objects up to higher positions, such as in the pyramids of the Middle East and Central America.

For example, to lift a barrel straight up to a height (*H*) requires a force equal to its weight (*W*). However, the force needed to lift the barrel is reduced by rolling it up a ramp, as shown below. So, if the ramp is *D* meters long and *H* meters high, the force (*F*) required to roll the weight (*W*) up the ramp is:

$$F = \frac{H}{D} \times W$$

Definition Sketch for a Ramp or Inclined Plane

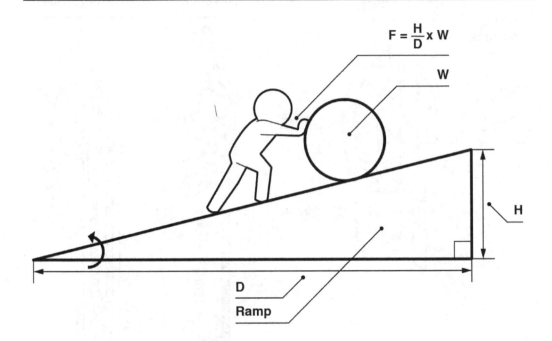

For a fixed height and weight, the longer the ramp, the less force must be applied. Remember, though, that the useful work done (in *N-m*) is the same in either case and is equal to *W* × *H*.

Wedges
If an incline or ramp is imagined as a right triangle like in the figure above, then a **wedge** would be formed by placing two inclines (ramps) back to back (or an isosceles triangle). A wedge is one of the six simple machines and is used to cut or split material. It does this by being driven for its full length into the material being cut. This material is then forced apart by a distance equal to the base of the wedge. Axes, chisels, and knives work on the same principle.

Screws
Screws are used in many applications, including vises and jacks. They are also used to fasten wood and other materials together. A screw is thought of as an inclined plane wrapped around a central cylinder. To visualize this, one can think of a barbershop pole, or cutting the shape of an incline (right triangle) out of a sheet of paper and wrapping it around a pencil (as in part *A* in the figure below). Threads are

made from steel by turning round stock on a lathe and slowly advancing a cutting tool (a wedge) along it, as shown in part *B*.

Definition Sketch for a Screw and Its Use in a Car Jack

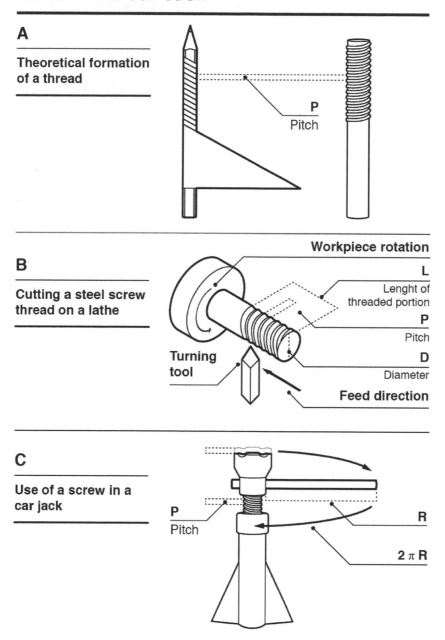

A

Theoretical formation of a thread

P
Pitch

B

Cutting a steel screw thread on a lathe

Workpiece rotation

L
Lenght of threaded portion

P
Pitch

D
Diameter

Turning tool

Feed direction

C

Use of a screw in a car jack

P
Pitch

R

$2 \pi R$

The application of a simple screw in a car jack is shown in part *C* in the figure above. The mechanical advantage of the jack is derived from the pitch of the screw winding. Turning the handle of the jack one revolution raises the screw by a height equal to the **screw pitch** (*p*). If the handle has a length *R*, the

distance the handle travels is equal to the circumference of the circle it traces out. The theoretical mechanical advantage of the jack's screw is:

$$MA = \frac{F}{L} = \frac{p}{2\pi R} \quad \text{so} \quad F = L \times \frac{p}{2\pi R}$$

For example, the theoretical force (*F*) required to lift a car with a mass (*L*) of 5000 kilograms, using a jack with a handle 30 centimeters long and a screw pitch of 0.5 cm, is given as:

$$F \cong 50{,}000 \; N \; \times \; \frac{0.5 \; cm}{6.284 \times 30 \; cm} \cong 130 \; N$$

The theoretical value of mechanical advantage doesn't account for friction, so the actual force needed to turn the handle is higher than calculated.

Practice Questions

The following Practice Test contains sample problems that reinforce the principles presented in the Mechanical Comprehension (MC) study guide. The answers to these problems, along with a brief explanation, follows.

1. A car is traveling at a constant velocity of 25 m/s. How long does it take the car to travel 45 kilometers in a straight line?
   a. 1 hour
   b. 3600 seconds
   c. 1800 seconds
   d. 900 seconds

2. A ship is traveling due east at a speed of 1 m/s against a current flowing due west at a speed of 0.5 m/s. How far has the ship travelled from its point of departure after two hours?
   a. 1.8 kilometers west of its point of departure
   b. 3.6 kilometers west of its point of departure
   c. 1.8 kilometers east of its point of departure
   d. 3.6 kilometers east of its point of departure

3. A car is driving along a straight stretch of highway at a constant speed of 60 km/hour when the driver slams the gas pedal to the floor, reaching a speed of 132 km/hour in 10 seconds. What's the average acceleration of the car after the engine is floored?
   a. 1 m/s^2
   b. 2 m/s^2
   c. 3 m/s^2
   d. 4 m/s^2

4. A spaceship with a mass of 100,000 kilograms is far away from any planet. To accelerate the craft at the rate of 0.5 m/sec^2, what is the rocket thrust?
   a. 98.1 N
   b. 25,000 N
   c. 50,000 N
   d. 75,000 N

5. The gravitational acceleration on Earth averages 9.81 m/s^2. An astronaut weighs 1962 N on Earth. The diameter of Earth is six times the diameter of its moon. What's the mass of the astronaut on Earth's moon?
   a. 100 kilograms
   b. 200 kilograms
   c. 300 kilograms
   d. 400 kilograms

6. A football is kicked so that it leaves the punter's toe at a horizontal angle of 45 degrees. Ignoring any spin or tumbling, at what point is the upward vertical velocity of the football at a maximum?

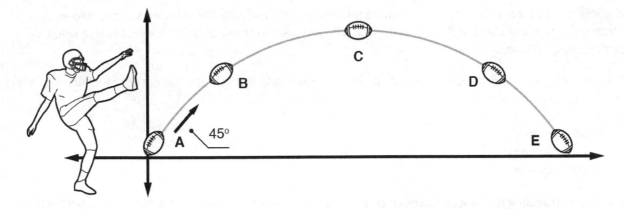

a. At Point A
b. At Point C
c. At Points B and D
d. At Points A and E

7. The skater is shown spinning in Figure (a), then bringing in her arms in Figure (b). Which sequence accurately describes what happens to her angular velocity?

(a)   (b)

a. Her angular velocity decreases from (a) to (b)
b. Her angular velocity doesn't change from (a) to (b)
c. Her angular velocity increases from (a) to (b)
d. It's not possible to determine what happens to her angular velocity if her weight is unknown.

8. A cannonball is dropped from a height of 10 meters off of the ground. What is its approximate velocity just before it hits the ground?

    a. 9.81 m/s

    b. 14 m/s

    c. 32 m/s

    d. It can't be determined without knowing the cannonball's mass

9. The pendulum is held at point A, and then released to swing to the right. At what point does the pendulum have the greatest kinetic energy?

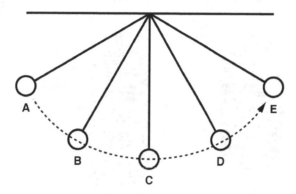

    a. At Point B

    b. At Point C

    c. At Point D

    d. At Point E

10. Which statement is true of the total energy of the pendulum?

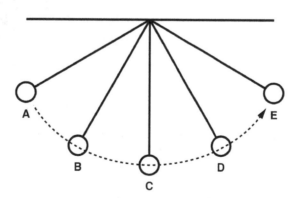

    a. Its total energy is at a maximum and equal at Points A and E.

    b. Its total energy is at a maximum at Point C.

    c. Its total energy is the same at Points A, B, C, D, and E.

    d. The total energy can't be determined without knowing the pendulum's mass.

11. How do you calculate the useful work performed in lifting a 10-kilogram weight from the ground to the top of a 2-meter ladder?
    a. 10kg x 2m x 32 m/s^2
    b. 10kg x 2m^2 x 9.81 m/s
    c. 10kg x 2m x 9.81m/s^2
    d. It can't be determined without knowing the ground elevation

12. A steel spring is loaded with a 10-newton weight and is stretched by 0.5 centimeters. What is the deflection if it's loaded with two 10-newton weights?

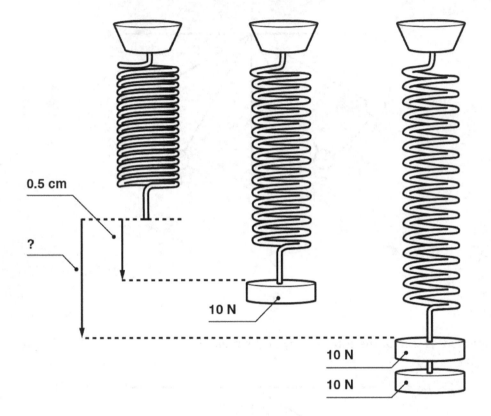

    a. 0.5 centimeter
    b. 1 centimeter
    c. 2 centimeters
    d. It can't be determined without knowing the Modulus of Elasticity of the steel.

13. A 1000-kilogram concrete block is resting on a wooden surface. Between these two materials, the coefficient of sliding friction is 0.4 and the coefficient of static friction is 0.5. How much more force is needed to get the block moving than to keep it moving?

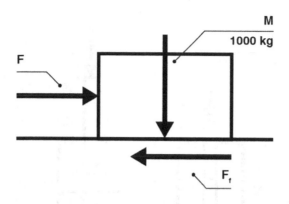

a. 981 N
b. 1962 N
c. 3924 N
d. 9810 N

14. The master cylinder (F1) of a hydraulic jack has a cross-sectional area of 0.1 m², and a force of 50 N is applied. What must the area of the drive cylinder (F2) be to support a weight of 800 N?

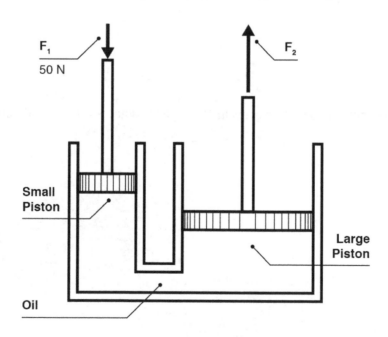

a. 0.4 m²
b. 0.8 m²
c. 1.6 m²
d. 3.2 m²

15. A gas with a volume V_1 is held down by a piston with a force of F newtons. The piston has an area of A. After heating the gas, it expands against the weight to a volume V_2. What was the work done?

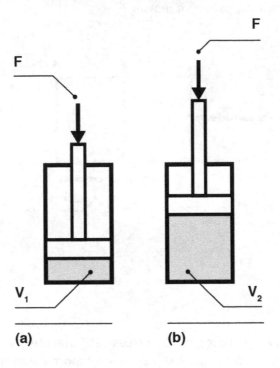

a. F/A
b. $(F/A) \times V_1$
c. $(F/A) \times V_2$
d. $(F/A) \times (V_2 - V_1)$

16. A 1000-kilogram weight is raised 30 meters in 10 minutes. What is the approximate power expended in the period?
a. $1000 \, Kg \times m/s^2$
b. $500 \, N \cdot m$
c. $500 \, J/s$
d. 100 watts

17. A 2-meter high, concrete block is submerged in a body of water 12 meters deep (as shown). Assuming that the water has a unit weight of 1000 N/m³, what is the pressure acting on the upper surface of the block?

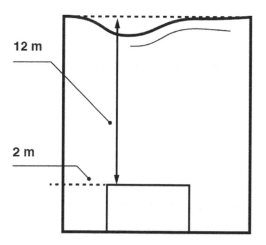

a. 10,000 Pa
b. 12,000 Pa
c. 14,000 Pa
d. It can't be calculated without knowing the top area of the block.

18. Closed Basins A and B each contain a 10,000-ton block of ice. The ice block in Basin A is floating in sea water. The ice block in Basin B is aground on a rock ledge (as shown). When all the ice melts, what happens to the water level in Basin A and Basin B?

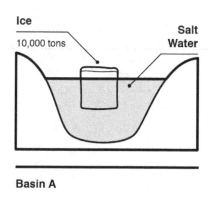

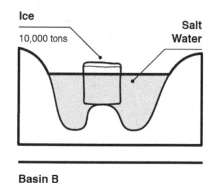

a. Water level rises in A but not in B
b. Water level rises in B but not in A
c. Water level rises in neither A nor B
d. Water level rises in both A and B

19. An official 10-lane Olympic pool is 50 meters wide by 25 meters long. How long does it take to fill the pool to the recommended depth of 3 meters using a pump with a 750 liter per second capacity?
    a. 2500 seconds
    b. 5000 seconds
    c. 10,000 seconds
    d. 100,000 seconds

20. Water is flowing in a rectangular canal 10 meters wide by 2 meters deep at a velocity of 3 m/s. The canal is half full. What is the flow rate?
    a. 30 m³/s
    b. 60 m³/s
    c. 90 m³/s
    d. 120 m³/s

21. A 150-kilogram mass is placed on the left side of the lever as shown. What force must be exerted on the right side (in the location shown by the arrow) to balance the weight of this mass?

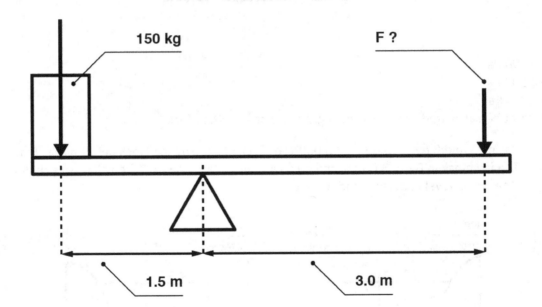

    a. 675 kg.m
    b. 737.75 N
    c. 1471.5 N
    d. 2207.25 N·m

22. For the wheel and axle assembly shown, the shaft radius is 20 millimeters and the wheel radius is 300 millimeters. What's the required effort to lift a 600 N load?

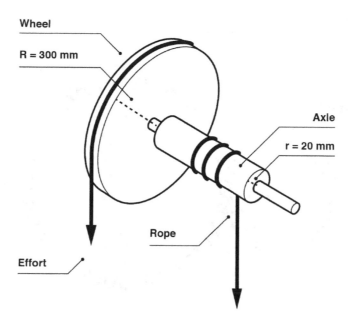

a. 10 N
b. 20 N
c. 30 N
d. 40 N

23. The driver gear (Gear A) turns clockwise at a rate of 60 RPM. In what direction does Gear B turn and at what rotational speed?

40 Teeth

B

20 Teeth

A

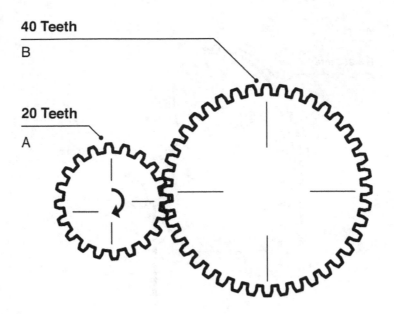

    a. Clockwise at 120 RPM
    b. Counterclockwise at 120 RPM
    c. Clockwise at 30 RPM
    d. Counterclockwise at 30 RPM

24. The three steel wheels shown are connected by rubber belts. The two wheels at the top have the same diameter, while the wheel below is twice their diameter. If the driver wheel at the upper left is turning clockwise at 60 RPM, at what speed and in which direction is the large bottom wheel turning?

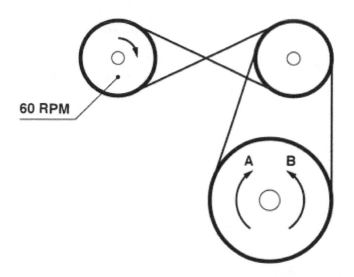

a. 30 RPM, clockwise (A)
b. 30 RPM, counterclockwise (B)
c. 120 RPM, clockwise (A)
d. 120 RPM, counterclockwise (B)

25. In case (a), both blocks are fixed. In case (b), the load is hung from a moveable block. Ignoring friction, what is the required force to move the blocks in both cases?

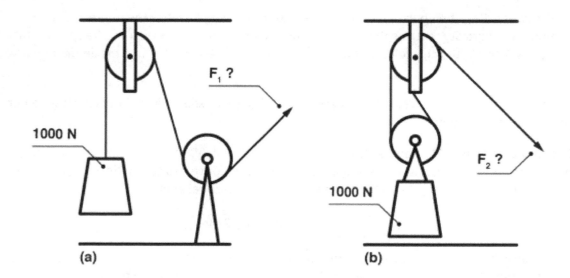

a. F_1 = 500 N; F_2 = 500 N
b. F_1 = 500 N; F_2 = 1000 N
c. F_1 = 1000 N; F_2 = 500 N
d. F_1 = 1000 N; F_2 = 1000 N

Answer Explanations

1. C: The answer is 1800 seconds:

$$(Desired\ Distance\ in\ km\ \times\ conversion\ factor\ (m\ to\ km))/current\ velocity\ in\ \frac{m}{s}$$

$$\left(45\ km\ \times\ \frac{1000\ m}{km}\right)\Big/25\frac{m}{s} = 1800\ seconds$$

2. D: The answer is 3.6 kilometers east of its point of departure. The ship is traveling faster than the current, so it will be east of the starting location. Its net forward velocity is 0.5 m/s which is 1.8 kilometers/hour, or 3.6 kilometers in two hours.

3. B: The answer is 2 m/s²:

$$a = \frac{\Delta v}{\Delta t} = \frac{132\frac{km}{hr} - 60\frac{km}{hr}}{10\ seconds}$$

$$\frac{70\frac{km}{hr} \times 1000\frac{m}{km} \times \frac{hour}{3600\ sec}}{10\ seconds} = 2\ m/s^2$$

4. C: The answer is 50,000 N. The equation $F = ma$ should be memorized. All of the values are given in the correct units (kilogram-meter-second) so just plug them in.

5. B: The answer is 200 kilograms. This is actually a trick question. The mass of the astronaut is the same everywhere (it is the weight that varies from planet to planet). The astronaut's mass in kilograms is calculated by dividing his weight on Earth by the acceleration of gravity on Earth: 1962/9.81 = 200.

6. A: The answer is that the upward velocity is at a maximum when it leaves the punter's toe. The acceleration due to gravity reduces the upward velocity every moment thereafter. The speed is the same at points A and E, but the velocity is different. At point E, the velocity has a maximum *negative* value.

7. C: The answer is her angular velocity increases from (a) to (b) as she pulls her arms in close to her body and reduces her moment of inertia.

8. B: The answer is 14 m/s. Remember that the cannonball at rest "y" meters off the ground has a potential energy of $PE = mgy$. As it falls, the potential energy is converted to kinetic energy until (at ground level) the kinetic energy is equal to the total original potential energy:

$$\frac{1}{2}mv^2 = mgy \text{ or } v = \sqrt{2gy}$$

This makes sense because all objects fall at the same rate, so the velocity *must* be independent of the mass (which is why "D" is incorrect). Plugging the values into the equation, the result is 14 m/s. Remember, the way to figure this quickly is to have $g = 10$ rather than 9.81.

9. B: The answer is at Point C, the bottom of the arc.

10. C: This question isn't difficult, but it must be read carefully:

A is wrong. Even though the total energy is at a maximum at Points A and E, it isn't equal at only those points. The total energy is the same at *all* points. *B* is wrong. The kinetic energy is at a maximum at C, but not the *total* energy. The correct answer is *C*. The total energy is conserved, so it's the same at *all* points on the arc. *D* is wrong. The motion of a pendulum is independent of the mass. Just like how all objects fall at the same rate, all pendulum bobs swing at the same rate, dependent on the length of the cord.

11. C: The answer is 10kg x 2m x 9.81m/s². This is easy, but it must also be read carefully. Choice *D* is incorrect because it isn't necessary to know the ground elevation. The potential energy is measured *with respect* to the ground and the ground (or datum elevation) can be set to any arbitrary value.

12. B: The answer is 1 centimeter. Remember that the force (*F*) required to stretch a spring a certain amount (*d*) is given by the equation *F* = *kd*. Therefore, *k* = *F*/*d* = 20N/0.5 cm = 20 N/cm. Doubling the weight to 20 N gives the deflection:

$$d = \frac{F}{k} = \frac{20N}{20N/cm} = 1 \; centimeter$$

All of the calculations can be bypassed by remembering that the relation between force and deflection is linear. This means that doubling the force doubles the deflection, as long as the spring isn't loaded past its elastic limit.

13. A: The answer is 981 N. The start-up and sliding friction forces are calculated in the same way: normal force (or weight) times the friction coefficient. The difference between the two coefficients is 0.1, so the difference in forces is 0.1 x 1000 x 9.81 = 981 N.

14. C: The answer is 1.6 m². The pressure created by the load is 50N/0.1m² = 500 N/m². This pressure acts throughout the jack, including the large cylinder. Force is pressure times area, so the area equals pressure divided by force or 800N/500N/m² = 1.6m².

15. D: The answer is (*F*/*A*) x (*V₂* -*V₁*). Remember that the work for a piston expanding is pressure multiplied by change in volume. Pressure = *F*/*A*. Change in volume is (*V₂* - *V₁*).

16. C: The answer is 500 J/s. Choice *A* is incorrect because kg x m/s² is an expression of force, not power. Choice *B* is incorrect because N·m is an expression of work, not power. That leaves Choices *C* and *D*, both of which are expressed in units of power: watts or joules/second. Using an approximate calculation (as suggested):

$$1000 \; kg \; \times \; 10\frac{m}{s^2} \times \; 30 \; m = 300,000 \; N \cdot m \quad so \quad \frac{300,000 \; N \cdot m}{600 \; seconds} = 500 \; watts = 500 \; J/s$$

17. B: The answer is 12,000 Pa. The top of the block is under 12 meters of water:

$$P = 1000\frac{N}{m^3} \times \; 12 \; meters = 12,000\frac{N}{m^2} = 12,000 \; Pa$$

There are two "red herrings" here: Choice *C* of 14,000 Pa is the pressure acting on the *bottom* of the block (perhaps through the sand on the bottom of the bay). Choice *D* (that it can't be calculated without knowing the top area of the block) is also incorrect. The top area is needed to calculate the total *force* acting on the top of the block, not the pressure.

18. B: The answer is that the water level rises in B but not in A. Why? Because ice is not as dense as water, so a given mass of water has more volume in a solid state than in a liquid state. Thus, it floats. As the mass of ice in Basin A melts, its volume (as a liquid) is reduced. In the end, the water level doesn't change. The ice in Basin B isn't floating. It's perched on high ground in the center of the basin. When it melts, water is added to the basin and the water level rises.

19. B: The answer is 5000 seconds. The volume is 3 x 25 x 50 = 3750 m³. The volume divided by the flow rate gives the time. Since the pump capacity is given in liters per second, it's easier to convert the volume to liters. One thousand liters equals a cubic meter:

$$Time = \frac{3,750,000\ liters}{750\ liters/second} = 5000\ seconds = 1.39\ hours$$

20. A: The answer is 30 m³/s. One of the few equations that must be memorized is $Q = vA$. The area of flow is 1m x 10m because only half the depth of the channel is full of water.

21. B: The answer is 737.75 N. This is a simple calculation:

$$\frac{9.81\ m}{s^2} \times 150\ kg \times 1.5\ m = 3\ m \times F \quad so \quad F = \frac{2207.25\ N \cdot m}{3\ meters}$$

22. D: The answer is 40 N. Use the equation $F = L \times r/R$. Note that for an axle with a given, set radius, the larger the radius of the wheel, the greater the mechanical advantage.

23. D: The answer is counterclockwise at 30 RPM. The driver gear is turning clockwise, and the gear meshed with it turns counter to it. Because of the 2:1 gear ratio, every revolution of the driver gear causes half a revolution of the follower.

24. B: The answer is 30 RPM, counterclockwise (B). While meshed gears rotate in different directions, wheels linked by a belt turn in the same direction. This is true unless the belt is twisted, in which case they rotate in opposite directions. So, the twisted link between the upper two wheels causes the right-hand wheel to turn counterclockwise, and the bigger wheel at the bottom also rotates counterclockwise. Since it's twice as large as the upper wheel, it rotates with half the RPMs.

25. C: The answer is F_1 = 1000 N; F_2 = 500 N. In case (a), the fixed wheels only serve to change direction. They offer no mechanical advantage because the lever arm on each side of the axle is the same. In case (b), the lower moveable block provides a 2:1 mechanical advantage. A quick method for calculating the mechanical advantage is to count the number of lines supporting the moving block (there are two in this question). Note that there are no moving blocks in case (a).

Assembling Objects

The **Assembling Objects (AO)** section tests a candidate's ability to think in three dimensions by having them follow a simple set of instructions and assemble an object from its component parts. This is an important skill because success in almost every technical field depends on being able to assemble complex systems from simpler parts or components. But relax! This is the same skill that's used to solve a jigsaw puzzle, fold origami, tie a knot, or take apart and reassemble simple machinery.

It's more important to become familiar with the test structure on the AO section than on any other part of the exam. Doing so will avoid wasting valuable time trying to understand the instructions when the AO section begins. To better understand this structure, think of a soldier reassembling a rifle or a technical specialist putting a pump back together. They start with the parts laid out to one side, then put the parts together according to the instructions. That's how the AO section is structured.

An example of a question similar to those found on the exam is shown below. Below the question are "bubbles" corresponding to the possible solutions shown in boxes 1 through 5. Choose the correct solution and fill in the "bubble" next to the letter that corresponds to it.

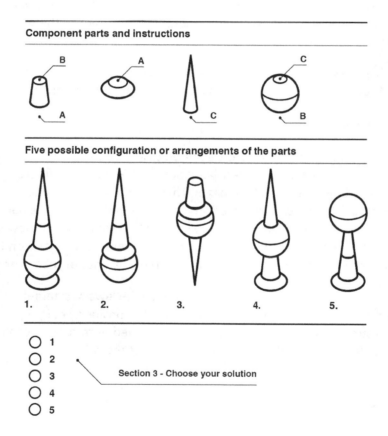

Understanding the Structure of an Assembling Objects (AO) Question

The most efficient way to solve AO questions is to systematically rule out the incorrect answers one at a time. Look at the entire system of parts first, and then work component by component.

In the example, the system consists of four components: a truncated cone marked *A* on the bottom and *B* on the top face; a shorter, squatter truncated cone labeled *A* on the top face; a thin, tipped cone marked *C* on the base; and a sphere with a line around its "equator" marked *C* on its "north pole" and a *B* on its "south pole." This is a simple system with four components connected "in series," which means one after the other.

Look at Box 1. The thin, tipped cone fits neatly onto the taller of the two truncated cones and sits on top of the sphere. However, the instructions show that the top of the truncated cones must be joined to *B*, and the sharp conic section must be joined at its base to *C*, so these two pieces don't fit together. Therefore, Box 1 is not the solution.

Box 2 can be rejected immediately because it shows the same incorrect assembly.

Now look at Box 3. The two truncated cones are properly connected, with the squat one serving as the base. However, the base is shown sitting on top of the sphere when the surface of a part labeled *C* should be sitting there (which is the sharp cone). In addition, the surface of the squat base is not connected to any part, so Box 3 isn't the answer.

Now that three possible answers have been eliminated, it's easy to compare the last two configurations (Box 4 and Box 5) and see that the only one matching the problem's assembly requirements is Box 4.

Preparing for the Test

First, it's important to understand what a correct solution to an assembly problem is. As mentioned before, the unassembled parts shown in the left-hand box can be moved to different positions, rotated, or even flipped (turned through the axis of the page). They can be turned around so not all of the parts are seen, like the way a coffee mug can be turned so the handle isn't visible. The parts can also be turned so that only one face is visible, like how a cube looks like a square when viewed head on, or a disk looks like a thin rectangle when viewed from the edge. However, the parts won't be stretched or bent, so think of them as being made out of solid wood, rather than out of rubber or plastic.

It's also important to remember that the rotations must obey the laws of symmetry. A right-handed object can't be turned so that it becomes a left-handed object. Consider a pair of gloves; a right-hand glove can be turned around or over but, no matter how it's turned or rotated, it can't be turned into a left-hand glove without turning it inside out—and that's against the rules.

Practice Questions

Section 1: For questions 1–12 below, which figure best shows how the objects in the left box will appear if they are fit together?

1.

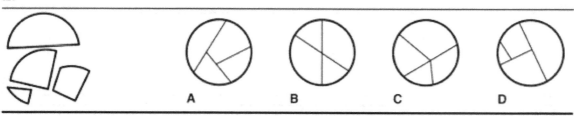

2.

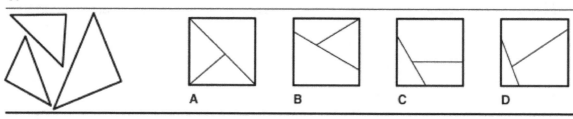

3.

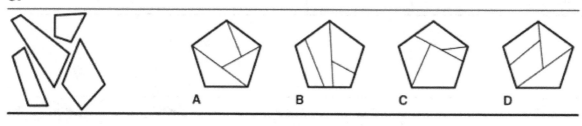

4.

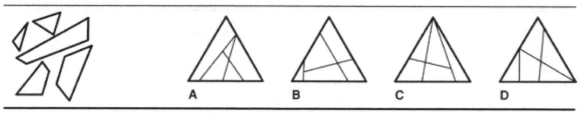

5.

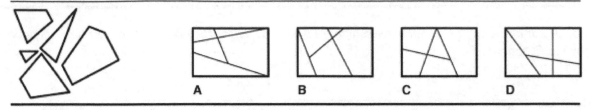

6.

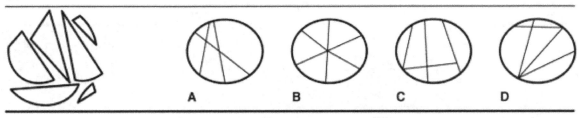

7.

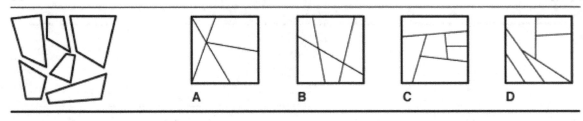

8.

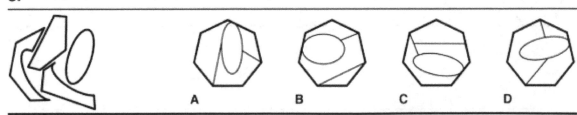

9.

A   B   C   D

10.

A   B   C   D

11.

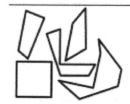   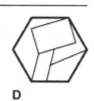

A   B   C   D

12.

A   B   C   D

Section 2: For questions 1–13 below, which figure best shows how the objects in the left box will touch if the letters for each object are matched?

1.

2.

3.

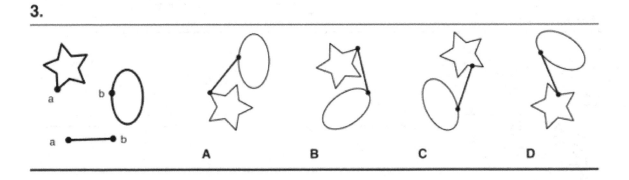

4.

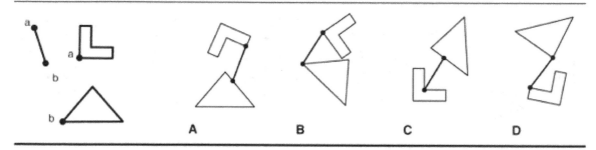

A   B   C   D

5.

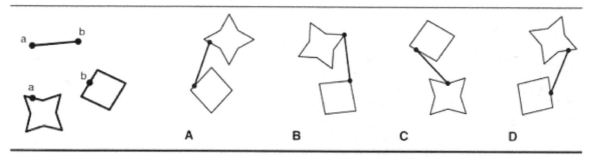

A   B   C   D

6.

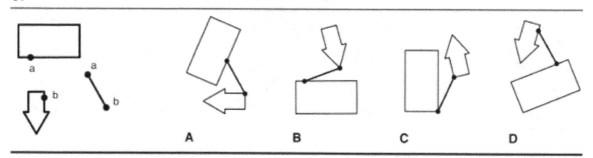

A   B   C   D

7.

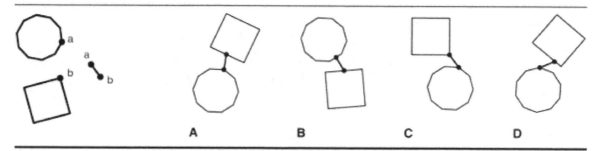

A   B   C   D

8.

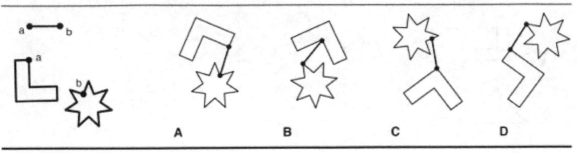

A B C D

9.

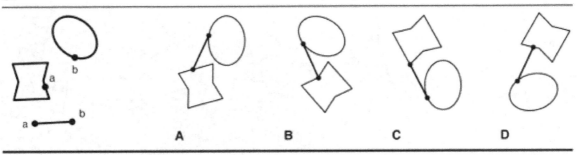

A B C D

10.

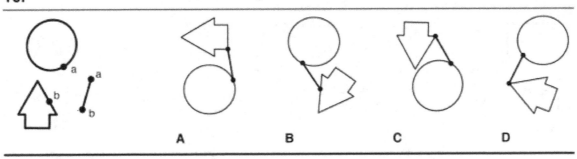

A B C D

11.

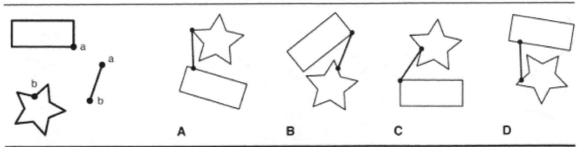

A B C D

12.

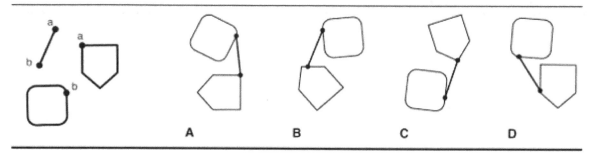

A   B   C   D

13.

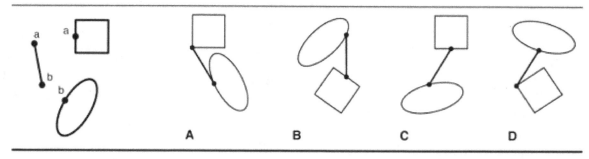

A   B   C   D

Answer Explanations

1. A
2. D
3. B
4. A
5. B
6. D
7. B
8. C
9. A
10. D
11. C
12. A

1. B
2. C
3. A
4. B
5. D
6. D
7. C
8. A
9. D
10. B
11. C
12. A
13. C

Dear ASVAB Test Taker,

We would like to start by thanking you for purchasing this study guide for your ASVAB exam. We hope that we exceeded your expectations.

Our goal in creating this study guide was to cover all of the topics that you will see on the test. We also strove to make our practice questions as similar as possible to what you will encounter on test day. With that being said, if you found something that you feel was not up to your standards, please send us an email and let us know.

We would also like to let you know about other books in our catalog that may interest you.

Test Name	Amazon Link
ASTB	amazon.com/dp/1628456809
AFOQT	amazon.com/dp/1628454776
OAR	amazon.com/dp/1628454482
SIFT	amazon.com/dp/1628454318

We have study guides in a wide variety of fields. If the one you are looking for isn't listed above, then try searching for it on Amazon or send us an email.

Thanks Again and Happy Testing!
Product Development Team
info@studyguideteam.com

Interested in buying more than 10 copies of our product? Contact us about bulk discounts:

bulkorders@studyguideteam.com

FREE Test Taking Tips DVD Offer

To help us better serve you, we have developed a Test Taking Tips DVD that we would like to give you for FREE. **This DVD covers world-class test taking tips that you can use to be even more successful when you are taking your test.**

All that we ask is that you email us your feedback about your study guide. Please let us know what you thought about it – whether that is good, bad or indifferent.

To get your **FREE Test Taking Tips DVD**, email freedvd@studyguideteam.com with "FREE DVD" in the subject line and the following information in the body of the email:

   a. The title of your study guide.

   b. Your product rating on a scale of 1-5, with 5 being the highest rating.

   c. Your feedback about the study guide. What did you think of it?

   d. Your full name and shipping address to send your free DVD.

If you have any questions or concerns, please don't hesitate to contact us at freedvd@studyguideteam.com.

Thanks again!

CPSIA information can be obtained
at www.ICGtesting.com
Printed in the USA
LVHW061309030820
662230LV00016B/1032

9 781628 458121